Building Decision Support Systems

Mark Wallace

Building Decision Support Systems

using MiniZinc

 Springer

Mark Wallace
Faculty of Information Technology
Monash University
Caulfield East, VIC
Australia

ISBN 978-3-030-41731-4   ISBN 978-3-030-41732-1   (eBook)
https://doi.org/10.1007/978-3-030-41732-1

Cover Figure: © Freepik.com

This Springer imprint is published by the registered company Springer Nature Switzerland AG.
The registered company address is: Gewerbestrasse 11, 6330 Cham, Switzerland

To Ingrid for all the lost evenings and weekends. . . Thank you for looking after absolutely everything while I just sat at my computer. I promise to join you back in real life now!

Preface

Thirty years ago, a number of researchers applied AI for the first time to optimisation. For applications from planning factory assembly lines to investment portfolios, the AI techniques proved more flexible and intelligible than traditional mathematical approaches. On the wave of the 4th industrial revolution, the application of AI to optimisation is now entering the main stream. AI is not just for machine learning but also, crucially, for decision support. AI can support the kinds of decisions made by professionals from many walks of life—managers in procurement, production, HR and finance, doctors, designers, captains and coaches.

This book meets three different kinds of objectives. For readers with an interest in the novel and changing ways in which computers affect our lives and livelihoods, this book provides an insight into intelligent decision support for optimisation and how computers can be made to do it. Chapters 1, 3 and Sect. 2.1 provide such an introduction; Sects. 5.1–5.3 introduce complexity; Sects. 10.1 and 12.1 introduce search and uncertainty, and Chap. 12 discusses the future.

For readers with a deeper interest in the power of computers and their limitations for solving optimisation problems, this book describes how different classes of problems are solved and the methods used by the computer to solve them. After the first 3 chapters, these readers can follow with Chaps. 5, 6 and 7 to understand which problems are hard to solve and why. Chapters 9, 11 and 12 cover search and uncertainty and a glimpse of the future.

Finally for readers who are interested in actually tackling such problems, the book introduces a modelling language and system that can be freely downloaded onto a home laptop and used to try solving problems for themselves. MiniZinc is applied to examples introduced in the other chapters of the book: Chap. 4 shows how to model in MiniZinc the problems of Chap. 2; Chap. 8 shows how MiniZinc can express problems of the different classes introduced in Chap. 6; and Chap. 10 covers search in MiniZinc. The MiniZinc chapters include exercises whose solutions are included in the appendix.

Scope of This Book

An intelligent decision support system for optimisation is a software tool to support enhanced decision-making. Its purpose is to enable the best choices to be made as measured against a set of goals. Intelligent decision support systems are used by a wide variety of people in a wide variety of application areas. Throughout this text, we will abbreviate 'Intelligent Decision Support' to IDS.

Decision support relies on access to information, based on correct and up-to-date data filtered, collated and summarised appropriately. It also relies on knowledge discovery and forecasting, enabling the user to recognise how different factors interact, and thus predict the consequences of different choices in the future.

The role of intelligent decision support, as covered in this text, is to exploit this knowledge and shape the future to best meet the user's goals. Our focus here is on combinations of choices. In a hospital, for example, an intelligent decision support system helps the management decide which operations will be carried out by which medical teams using which facilities over a period of time. The challenge is to balance choices against each other. When designing an aicraft, capacity, comfort and functionality must be balanced against weight, cost and reliability. Puzzles, such as Sudoku, are entertaining examples of making combinations of choices that fit together.

As will become clear, the interaction between the choices means that there are a huge number of possible combinations. For example, in a Sudoku puzzle there are (at most) 81 decisions to make—i.e. numbers to be filled in—and each decision has 9 choices. However, the number of correct Sudoku grids is (surprisingly) much more than 81 times 9. (In fact, the number is 6,670,903,752,021,072,936,960.) Finding the best combination of choices is consequently a very challenging problem.

This book is designed to enable you:

- to understand the potential benefits of deploying an IDS system
- to recognise the key risks in implementing an IDS system and know which techniques can be applied to minimise them
- to understand the technology of decision support at sufficient depth to manage or monitor an IDS project
- to distinguish good sense from mere jargon when dealing with everyone involved in an IDS project, from sales personnel to software implementers

As we shall see, the word 'intelligent' in IDS signifies that this technology is not (yet) standard, encapsulated in off-the-shelf modules which can be simply plugged together. Consequently, to manage an IDS project it is necessary to have an insight into the design and behaviour of each component. Therefore, the text addresses a number of technical issues:

- the nature of 'combinatorial' problems
- the distinction between 'modelling' a problem and 'solving' it
- the main algorithms for solving decision support problems

- the strengths and weaknesses of the different algorithms
- different ways of achieving scalability in a solution

Finally, this text does not cover the topics of 'big data' or 'deep learning', which are already well covered in the current literature.

Hands-on Learning

To support the learning process, the reader is encouraged to try out the ideas described in the text on toy applications and puzzles. This will help the reader

- learn to model problems
- learn to recognise shortcuts
- learn to exploit intelligent solving techniques

The IDS tool to be used for these experiments is a modelling language and implementation in which problem modelling and problem solving are kept separate. It provides

- a high-level modelling language
- interfaces to multiple solvers
- control over the search for solutions

This software is called 'MiniZinc' and is freely available to download from the site:

```
www.minizinc.org
```

This site includes software, tutorial and reference manual.
The chapters presenting MiniZinc all include challenges and exercises.

However, the book can be read to the end without the MiniZinc chapters: all the concepts are explained without reference to MiniZinc.

Acknowledgements

I am deeply grateful to Monash University for giving me the time to complete this book and to our company Opturion for the opportunity to put these ideas into practice in industry.

Peter Stuckey and Kim Marriott made it possible for me to join the fertile research community in Melbourne. Thank you.

Many thanks to Guido Tack, and his team, who have made MiniZinc such a wonderful tool for optimisation and decision support.

Finally, thanks to the many friends and colleagues who have given me feedback on draft chapters. From Monash: Rejitha Ravindra, Daniel Guimarans, Ilankaikone Senthooran, Arthur Maheo, Aldeida Aleti, Kevin Leo and Gleb Belov. And from my family: Bob Douglas, Tessa Wallace and Ingrid de Neve—thank you not just for letting me bring my work home but for agreeing to do it for me!

Caulfield East, VIC, Australia   Mark Wallace

Contents

Chapter 1
Motivation for IDS

The topic of this book is intelligent decision support, or *IDS*. This first chapter reviews how, over the years, the main users of IDS have risen up the corporate hierarchy. The meaning of "intelligence" in this context is discussed, and the risks of this kind of intelligence. The chapter concludes with a survey of the wide scope of IDS in strategic, tactical and operational applications.

1.1 History

IDS belongs to a new generation of IT uses, each generation addressing a greater level of complexity than the one before.

1.1.1 Menial Tasks

The first generation of IT addressed tasks which we can broadly classify as "menial". Menial tasks are ones which are boring, and unskilled. The menial tasks taken over by IT systems in the 19th and early 20th century were repetitive tasks that previously required people a very long time to carry out.

The IT systems that were designed to take over these tasks were specialised machines designed for the specific tasks. For example the Jacquard loom, invented in 1801, was able to automatically weave a pattern "programmed" by punched wooden slats (we would now call them "punched cards") sequenced along a pair of strings. Punched cards were also used by Hollerith for the 1890 census in the

© Springer Nature Switzerland AG 2020
M. Wallace, *Building Decision Support Systems*,
https://doi.org/10.1007/978-3-030-41732-1_1

USA. A census was taken of the US population every 10 years, from 1790. The data from the 1790 census had taken 9 months to process, but the 1880 census had taken 9 years, and with the growth in population it was clear that census data processing would have to be dramatically speeded up if it was to keep up with the censuses! Using punched cards to represent the data enabled it to be quickly sorted by different properties (combinations of holes)—another repetitive job that could be speeded up by using IT [18, p. 32].

The first calculating devices were also machines, such as Leibniz's "stepped reckoner" for addition, subtraction, multiplication and division built in 1674, and Babbage's difference engine designed in 1849. Again their uses were mundane—Schulz' difference engine, based on Babbage's design, was used for creating tables of logarithms.

Indeed many modern uses of computers are of this mundane kind: sorting, counting, and data processing. By now computers are cheap enough to be used for almost all repetitive tasks which arise in organisations: payroll calculation, order processing, billing, stock control, and so on.

1.1.2 Managerial Tasks

The next generation of tasks taken over by IT systems require more sophistication: more precise and up-to-date data, more complex processing, and additionally much more sophisticated user interfaces. These tasks are ones that would previously have been carried out by more senior members of an organisation—managers. The core function of these tasks is to provide the right information, at the right place at the right time (a phrase that computer sales people used for many years). The technology supporting this functionality includes databases and powerful data retrieval facilities; communication and distributed computing and, of course, the internet; and flexible easy-to-read user interfaces that quickly reveal patterns and exceptional items in the data.

These tasks are supported by "Business Intelligence" tools that are widely available today. However the name "Business Intelligence" is arguably inappropriate for two reasons:

1. These tasks (and tools) are applicable to all organisations, in commerce, government, schools, hospitals, defence and so on. For example a typical application is the monitoring of traffic volumes in urban road networks: at the control centre will be a screen showing, for example, where excessive traffic congestion is building up, where incidents have occurred, and at what point normal traffic flow has been restored.

2. "Intelligence" is used above in the (military) sense of "information" rather than its usual informal sense of "cleverness". However in the current text we will be using the word in a sense much closer to its informal meaning. We will shortly discuss the exact meaning of the word "Intelligent" in the title "Intelligent Decision Support Systems".

1.1.3 Entrepreneurial Tasks

An entrepreneurial task is an application of IT that differentiates an organisation from its competitors. It requires an IT system that is, in some sense, unique and ahead of the pack. The term also applies to IT applications in government organisations that make them stand out such as the European Medium Range weather forecast model, and UK National Health online.

Naturally if an application works well, then it quickly spreads, and eventually becomes encapsulated as a standard IT tool used by all the (competing) organisations. Consequently today's entrepreneurial task is, tomorrow, simply a managerial task.

We can look back at the recent history of IT to see a sequence of applications, such as Word processing, spreadsheets, email, online auctions, and social networking, that started as entrepreneurial applications but are now standard IT tools.

Entrepreneurial IT can not only enable existing tasks to be performed more efficiently, but they can create completely new business models. A classic example is the use of IT in logistics which enabled Fed-Ex to emerge from an idea written up as an assignment by Fred Smith at Yale University in 1965 (it was graded 'C'), to the launch of the business in 1973. By the year 2000 Fed-Ex had a cargo fleet of 640 aircraft and 38,500 trucks and a value of $16 billion.

1.2 "Intelligent" in IDS

1.2.1 The Meaning of "Intelligent" in IDS

As IT tasks progress from menial, to managerial to entrepreneurial, we see the users of IT climbing up the corporate ladder. The menial tasks are those taken over from junior employees who, at least in in certain organisations, "are not payed to think". The information processing involved in a database is considerable, as anyone who has taken a database course can vouch. Similarly spreadsheets, electronic calendars, and project management software require a great deal of clever information processing, to compute the value of a cell, the best date for a meeting or the earliest completion date of a task.

However there is nothing magic about these applications. We may not understand exactly how a spreadsheet macro computes the value of a cell, but we are accustomed to computers evaluating functions and to us a spreadsheet is just a tool, as a slide rule and abacus were to their users.

Intelligent is a word that we apply to people who can do something clever: it is the capability to think or accomplish clever things. In IDS we use "intelligent" for a (decision support) capability which would encourage us to call a person "intelligent" if they had it.

Examples of IDS systems are, accordingly, systems that can help us do something clever like planning a complex operation, optimising a financial portfolio or managing a set of resources so that all the required tasks can be completed in time.

Interestingly, when applied to an IT system, this meaning of "intelligent" is likely to change over time, just like the set of tasks which we might call "entrepreneurial". In the 1970's a calculator was considered a very clever gadget, and many people might have called it "intelligent". By now, however, people familiar with calculators perceive them as simple. Indeed packing any amount of (known) functionality within a small instrument—such as the latest mobile phone/tablet/netbook combination—would not nowadays make the system count as "intelligent".

By contrast, the defeat of a chess grandmaster by a computer (Deep Blue) in 1997 was a major event. In fact Kasparov himself saw "deep intelligence and creativity in the machine's moves".

When we use the word "intelligent" in IDS, or in describing any computer system, we are not claiming that computers are in any way like humans, nor that they have consciousness, and nor even that they think. Rather the word "intelligent" applied to a computer system means:

- that the computer can do tasks which would encourage us to call a person "intelligent" who was capable of doing such tasks
- that the computer is accomplishing tasks complicated enough that the present day observer cannot see how they are done
- that the computer is doing things that are currently clever IT

The first of these criteria is the hardest to satisfy, the others yielding more and more widely applicable uses of the word "intelligent".

1.3   Risks in "Intelligent" IDS

When IT systems are described as "intelligent" there is additionally a dark side, reflected in many science fiction films in which such systems have taken over control, and begun to serve their own objectives. These scenarios are often far-fetched, but there are good reasons to be cautious in dealing with such systems.

The risks arise because it is not possible to fully specify an IDS system—this is essentially why we call such systems "intelligent". Consider the chess playing system, Deep Blue, mentioned earlier. How would a chess playing IDSS be specified:

- it must obey the rules of chess
- it must make choices leading towards a winning position—again as given by the rules of chess

However the key aspect of decision-making cannot be easily specified (or chess playing would be as easy as noughts-and-crosses):

- what is a good position?
- how far ahead should I look to ensure the opponent cannot take my pieces?

Similarly in specifying a decision support system for a bank loans application: it is relatively straightforward to specify the rules set down by the management, but almost impossible to specify good lending behaviour: arguably it takes an indefinable quality called "common sense" to react to certain novel situations.

Here are two examples to illustrate the risks associated with intelligent IT.

- The publication of the Black-Scholes equation made it possible to value futures, and even deploy intelligent computer systems to trade in these financial instruments. It may well have been the fact that many of these automated systems employed the same criteria for buying and selling, that led to the market crash in 1987. At the end of 2004, the total amount of derivatives contracts outstanding worldwide summed to nearly $50K for every man, woman and child on earth! At least one of the causes of the global financial crisis of 2007–10 was that few people understood the web of debt in a system whose complexity was fuelled by IDS
- Many vehicles now offer computer-aided driving support, from Automated Braking Systems (ABS) to parking control, and detectors to monitor car positioning and driver wakefulness. Driverless vehicles are now starting to arrive on public roads, and have already been involved in fatalities. The risk for driverless vehicles is that new situations arise every day, and the software cannot be designed and tested for every possible eventuality.

One risk of decision support systems is that their users can become less involved with the decision-making. If the systems makes a wrong decision rarely enough, its users start to simply assume the computer will always generate the right decision. Unfortunately the very intelligence of IDS systems makes them unpredictable, and in a new situation their correct—and safe—behaviour cannot be guaranteed. Indeed when such systems err, they often go badly wrong, and human intervention is then critical.

1.4   IDS Scope

The decisions made by and for an organisation vary in their scope and impact. Certain *strategic* decisions set the direction of the organisation. Given a certain direction, a set of *tactical* decisions are involved in planning the organisation's activities. Finally when the planned activities are carried out, many *operational* decisions are required to keep things on track.

In effect the scope of strategic decisions are the widest. Tactical decisions have their scope limited by the strategic decisions. Finally the scope of each operational decision is dictated by the tactical decisions which preceded it.

In this section we will explore how IDS applies to decisions with these different scopes.

1.4.1   Strategic IDS

Strategic decisions are the major decisions which dictate the future of an organisation. They are made with a view to the longer-term future from 5 years and upwards. Examples of such decisions are what kinds of power plants a government should invest in, where to build motorways, where to locate factories, and to which countries an airline should run flights.

These decisions are characterised by the huge commitments they involve:

- money
- time
- manpower and resources

Their consequences are also dramatic: the right strategic decisions can make a company, and the wrong ones can break it. In particular the right commitment can create further growth opportunities, as well as increasing profitability and company reputation.

On the other hand strategic decisions are challenging because:

- the number and nature of the options may be unclear
- the commitments required as a result of each choice may be imprecisely understood
- the payoffs from each choice may also be unclear in advance

The methods covered in this text cannot eliminate these uncertainties. The commitments and payoffs may depend on the weather or the price of oil. The options available may depend on domestic voters or foreign powers!

Nevertheless IDS can support strategic decision making by analysing and quantifying the cost of implementing each choice. For example a decision concerning investment in power plants can be analysed in terms of the operation of different possible combinations of power plants under different scenarios. For hydro-power the likelihood of drought in different years and different times of the year must be weighed against the financial and environmental cost of other energy sources with the goal of meeting total power demand at all times of the year and yet minimising cost and environmental impact.

1.4.2   Tactical IDS

Tactical decision-making is the core application of IDSS. Tactical decisions address issues on a timescale of weeks and months, up to a few years. Typically they involve resourcing, in its many different guises. Typical examples of tactical decision-making are scheduling classes in a school or university, managing investments, transport scheduling, production planning and ordering raw materials, logistics, personnel rostering, resource allocation and budgeting.

The impact of tactical decisions are typically less dramatic than strategic decisions, but they are still crucial for the efficient running of an organisation, affecting both profitability and reputation. Indeed most tactical decisions seek both to optimise the service provided, and to minimise the cost of providing that service.

Tactical decision support is often easier than strategic decision support because

- the decisions involve known resources, or a known budget
- the commitments resulting from each choice are reasonably clear
- the benefits resulting from each choice are largely predictable
- the constraints governing the decisions are understood

1.4.3   Operational IDS

Operational decision-making is required to maintain efficiency during operation. One example is the choice of resource to allocate to tasks arising during the day of operation. For example a car breakdown company must decide which patrol to send to each breakdown, or a network provider allocating engineers to clients' home visits. Often some of the tasks are known in advance, while others arise during the day. Another kind of operational decision making is required when disruptions occur—due to jobs lasting longer than expected, last minute tasks, or late deliveries.

Sometimes operational decisions can be handled by simple rules such as sending the nearest engineer to a client when a call comes in. However the decisions are often complicated by deadlines, and the fact that resources are currently working on other tasks when a new request comes in.

The timescales for operational decision-making are typically short—from a few milliseconds to a few days. Usually the resources available are pretty much fixed and the goal is to handle as many tasks as possible with the resources available. The impact of operational decisions is limited: naturally operational decisions can make a big difference to a few clients, but the impact on the organisation as a whole is less significant than that of strategic and tactical decisions. The main benefits are to maintain the good reputation of the organisation and keep the clients satisfied.

Operational decision support is generally less complex than tactical decision support because

- the resources are largely fixed
- the range of choices available are limited, and highly constrained
- in cases where tasks arise during operation, the allocation of resources cannot be planned in advance. They must be simply chosen on the spot

1.5 Summary

Intelligence underpins the value of IDSS and its particular challenge as a technology. Systems that can be built from off-the-shelf software components are of great commercial importance, but we would not call them "intelligent". Intelligent systems can also be built from components, but their very combination depends on the implementation of the components themselves.

Decision support can be used for long-term strategic planning, medium term tactical scheduling and even online control. However when decision support becomes automated decision-making, unexpected outcomes occur—such as the 1987 market crash.

More generally the deployment of an IDSS must be undertaken intelligently. An IDSS

- Cannot be deployed without understanding
- Cannot be marketed without understanding
- Cannot be used without understanding

In other words Intelligent systems need

- Intelligent developers
- Intelligent marketing and integration
- Intelligent users

Chapter 2
Modelling and Choices

When a client seeks an IDSS to help with their problem, then one might assume they could at least say what the problem is. However as is often the case with IT systems in practice, developing a clear, correct and precise problem statement is actually a significant part of solving the problem. The other part of solving the problem is developing an algorithm that can take the problem data as input and return an optimal answer, or at least some good answers.

In this chapter we will use some simple example problems to illustrate what it means to "model" a problem [65].

2.1 Some Example Problems and Their Solutions

2.1.1 Selecting Numbers from a Set

We start with a couple of variations on a simple problem in order to gain an intuition about what makes a problem difficult to answer. The data is simply a set of numbers, say {7, 10, 23, 13, 4, 16}. We now present a sequence of problems involving the numbers. None of them are very hard problems (after the entire set has only six numbers), but the idea is that each problem gets a bit harder.

1. Find the largest number.
   The answer is, of course, 23. One way we can find this is by looking through the numbers one after another, remembering the biggest one we have seen so far, and replacing it the next one if it is larger.

© Springer Nature Switzerland AG 2020
M. Wallace, *Building Decision Support Systems*,
https://doi.org/10.1007/978-3-030-41732-1_2

2. Find the two largest numbers.

   This can be solved by finding the largest number, removing it, and finding the largest among the remainder. More cleverly we could go through the numbers one at a time, remembering the *two* largest numbers, and updating them if the new number is larger than either of them.

3. Are there four numbers with a total over 40?

   It is easy to find an answer—simply select the first four numbers. However this was just lucky.

4. Are there four numbers with total over 63?

   This is not quite so easy. However if we find the four largest numbers (by a method similar to the second problem), we can then add them, to give a total of 62. Since these are the largest four numbers there cannot be another combination of four numbers with a larger total. The answer is therefore no.

The last two problems are the same, except for the total number sought—40 in problem 3 and 63 in problem 4. Indeed we could generate many different problems by choosing different values for this total. For each value of the total we have a different problem   *instance*. Each of these problem instances is an instance of one problem   *class*. The problem class asks, for any given number K, are there four numbers with a total over K. K is termed the *parameter* of the problem class. Different problem instances are created by choosing different values for the parameter.

We can generalise further by making the count of numbers we add to make the total another parameter N: "Are there N numbers in $\{7, 10, 23, 13, 4, 16\}$ with a total over K?" This problem class has two parameters: the count and the total.

Indeed we can generalise further and introduce a third parameter, S, the set of numbers[1] from which we are choosing N numbers. The new problem class asks: "Are there N numbers in the set S with total over K?" Every instantiation of the parameters forms another problem instance. In particular instantiating S to $\{7, 10, 23, 13, 4, 16\}$, N to 4 and K to 40 gives problem (instance) 3 above.

It is possible to specify an algorithm to solve any problem in this most general problem class.

1. If the set S has fewer than N elements, then the answer is no.
2. Otherwise, repeat N times:

   a. find the maximum number in S
   b. remove it from S and record it

3. add the recorded numbers together
4. If their sum is greater than K, then answer yes
5. Otherwise, answer no

---

[1]If we want to allow the same number to occur more than once we can call it a "multi-set".

This algorithm is not itself important—and indeed one could refine it to solve the same problem class more efficiently. However it is an illustration of a generic algorithm applicable, not just to a single problem instance, but to a whole class of problems.

Surprisingly for a very similar problem class "Are there N numbers in the set S with total equal to K?" the most efficient algorithm is much more complicated and less efficient. (Try finding a set of 4 numbers from $\{7, 10, 23, 13, 4, 16\}$ with total equal to 41!) However, this is the subject of the next section.

2.1.2   The Knapsack Problem

The knapsack problem is the first of four problem classes that will be used throughout the next few chapters to illustrate how to set about problem modelling and problem solving.

The knapsack problem is this: Given a set of items, each with a weight and a value, determine how many of each item to include (in the knapsack) so that the total weight is less than or equal to a given limit and the total value is as large as possible.

While it seems trivial, this is a fundamental problem in logistics: whether loading ships, trucks, aircraft—or even the space shuttle!

There are many variations on the knapsack problem. In the "decision" variant, instead of optimising the total value, it is simply required to find a solution with, or exceeding, a certain value.

Secondly instead of deciding the number of each item to include, the choice might be to include the item (once) or not. This is the "0–1" version of the knapsack problem.

In a third variant, there is more than one limit on the items included. For example each item has a volume as well as a weight, and there is a limit on the total volume as well as a limit on the total weight. This is the "multiple constraint" knapsack problem.

There are other variations as well, but they need not concern us here (Fig. 2.1).

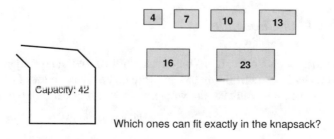

Which ones can fit exactly in the knapsack?

Fig. 2.1 The knapsack problem

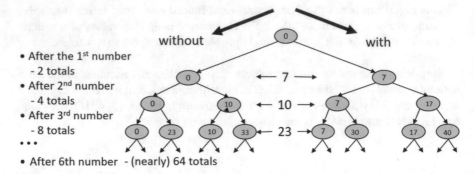

- After the 1st number
  - 2 totals
- After 2nd number
  - 4 totals
- After 3rd number
  - 8 totals
- • •
- After 6th number - (nearly) 64 totals

Fig. 2.2 All subsets and their subtotals

Our example from the previous section is a decision version of the 0–1 knapsack problem, but in this case we allow any number of different items to be added to the total—not a specific number N of items.

- Each item can be chosen once, or not
- The value of an item is always equal to its weight
- We will not optimise the value of included items, but instead try to find a solution whose total value is equal to the capacity of the knapsack (it cannot be greater, as it would then violate the limit).

This version of the problem asks if S is a set of integers, and K is the target value, is there a subset of S whose sum is K.[2] (A *subset* of a set contains just some of the elements of the set.)

Given a particular set (say {7, 10, 23, 13, 4, 16}) and a particular total (say 41), we can find the answer by listing the numbers in order: 7, 10, etc., and then either adding, or not adding, each to the total. This computes all possible totals:

A simple algorithm to find a subset with the required total is to generate this whole tree and see if any of the leaf-nodes (in the final row of the tree) have the right total. However there is no apparent way to generate the leaves of the tree in a certain order, so as to stop early if there is no node with the right total (Fig. 2.2).

2.1.3   Breaking Chocolate

You have a block of chocolate. It is a big block, with 13 rows, and 9 squares in each row, so it contains 117 squares! You task is to break the block up into 117 individual squares. At the first step you take the whole block and break it into two pieces. At

---

[2]This is also called the subset-sum problem.

each subsequent step you take just one piece, and break it into two. When you break a piece into two you can only snap it along either a vertical or a horizontal line.

The problem is to break the block into individual squares *in the possible least number of steps*. How many steps do you need?

This problem seems difficult. At each step there is a choice of which piece to break next. Having chosen a piece, the choice is whether to break it on a vertical or horizontal line. Having chosen that, the question is which line to break it down—a line nearer the edge or nearer the middle? Consider how you might try and solve the problem. The answer is at the end of this chapter.

2.2 The Basics of Problem Modelling

2.2.1 Fundamental Concepts and Definitions

In this section we will develop the fundamentals of problem modelling. Specifically we will introduce the set of concepts used in constructing models that can be processed by an IDSS. The three concepts—the basic building blocks of an IDSS model—are:

Variables   representing the decisions; a value for a variable represents a *choice* for that decision.
Constraints   ruling out incompatible choices
Parameters   specifying a problem instance

A constraint is imposed on a set of variables, called the variables in its *scope*, and restricts the possible combinations of values taken by those variables. For example the constraint $X > Y$ on the variables X and Y, rules out any combination in which the value of X is not greater than the value of Y. The scope of $X > Y$ is the set of variables it constrains, $\{X, Y\}$.

An assignment of values to the variables in the scope of a constraint, can either *satisfy* the constraint or *violate* it. The assignment $X = 1, Y = 2$, of value 1 to variable X and value 2 to variable Y violates the constraint $X > Y$. On the other hand the assignment $X = 2, Y = 1$ satisfies it.

When modelling problems it is useful to specify the range of choices for each decision: indeed for the decision-maker to be able to consider all the possible choices, there can only be a finite number of choices for each decision.

Therefore we specify for each (decision) variable a set of possible (choices) values, which we call the *domain* of the variable.

The purpose of an IDSS is to support the making of a set of decisions, represented in terms of the model by assigning values to variables. A *feasible assignment* is an assignment over a set of variables that satisfies every constraint whose scope is a subset of this set. A feasible complete assignment therefore satisfies all the constraints in the problem model!

A *complete assignment* is an assignment, to each variable in the model, of a value in its domain.

A *solution* is a feasible complete assignment.

2.2.2   Definitions of Modelling Concepts

Definition: A **model** is the representation of a problem in terms of its variable, its constraints and its parameters.[3]

Definition: A **variable** represents something unknown when the model is defined. It is assigned a specific value when the model is solved.

Definition: A **domain** is the set of alternative values that a variable could take.

Definition: An **assignment** to a set of variables associates with each variable a specific value from its domain. A **complete assignment** assigns a value to every variable in a model

Definition: A **constraint** specifies conditions on the variables in its scope. When the model is solved, the values assigned to its variables must satisfy the constraint

Definition: A **feasible assignment** is an assignment over a set of variables that satisfies every constraint whose scope is a subset of this set.

Definition: A **solution** is a feasible complete assignment.

Definition: A **parameter** is a value that is input to a model.

Definition: A **problem class** is a general problem. The same model can be used for all the problems in a problem class, but one or more of the inputs are as yet unspecified.

Definition: A **problem instance** is a specific problem. All the inputs for a problem instance are already specified.

Definition: An **array** is a fixed-length list of values, or variables.[4]

Definition: A **matrix** is a table, or more generally, a multi-dimensional array of values or variables.

Definition: An **operation** on a list of variables and/or values returns a **result** which is a value. The list of values or variables are termed **arguments** of the operation. If they are all values they are often termed "inputs".

---

[3]Later we will also include its **objective**.

[4]Later it will be shown that they must be of the same **type**.

2.2.3   Modelling Some Example Problems

We will illustrate the use of these concepts in modelling some example problems.

Boats and Docks

This is essentially a harbourmaster's problem. When a number of boats are waiting to enter harbour, to allocate a dock to each one. Since there are not usually enough docks to accommodate all the boats at once, several boats might need to use the same dock (perhaps one after another).

A typical constraint would be that certain docks are too small to fit certain boats, or certain boats cannot wait, but for the purposes of this illustration we do not need to know all the constraints of the problem (Fig. 2.3).

Assume there are three docks, *dock1*, *dock2* and *dock3*, and five boats *b1, b2, b3, b4* and *b5*. The dock size constraint says that boats *b1* and *b2* cannot fit in *dock1*.

We can model this problem with a variable for each boat representing the decision: to which dock should the boat be assigned? *Boat1, Boat2, Boat3, Boat4, Boat5*. Possible values for each variable are the docks to which they can be allocated. Thus the domain of *Boat1* is {*dock2, dock3*}. *Boat2* also has the domain {*dock2, dock3*}. The other boats have domain {*dock1, dock2, dock3*}.

A partial assignment is: *Boat1 = dock2, Boat3 = dock1*. A complete assignment is *Boat1 = dock2, Boat2 = dock3, Boat3 = dock1, Boat4 = dock2, Boat5 = dock1*. If the above complete assignment satisfies all the constraints of the problem (which we have not stated), then it is a feasible complete assignment, and therefore a solution.

Often a problem has many solutions, though only some of them may be wanted. For example we are often only looking for the "best" solution or solutions.

Definition A *candidate* solution is a complete feasible assignment. It may, or may not, be the desired solution.   □

Fig. 2.3 Allocating boats to docks

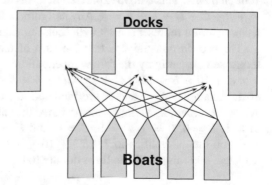

It is worth counting the candidate solutions for this problem. The variable *Boat1* can take two possible values, and so can *Boat2*. Consequently we can list four partial assignments over *Boat1*, *Boat2*:

Boat1 = *dock2*, *Boat2* = *dock2*
Boat1 = *dock2*, *Boat2* = *dock3*
Boat1 = *dock3*, *Boat2* = *dock2*
Boat1 = *dock3*, *Boat2* = *dock3*

Each of these four partial solutions can be extended in three different ways to partial solutions over *Boat1*, *Boat2*, *Boat3* by assigning *Boat3* = *dock1* or *Boat3* = *dock2* or *Boat3* = *dock3*, making a total of 12 different partial solutions over *Boat1*, *Boat2*, *Boat3*. The number of complete assignments (otherwise known as candidate solutions) is $12 \times 3 \times 3 = 108$, since each of the variables *Boat4* and *Boat5* can also be assigned to one of three different values. This number, 108 is $2 \times 2 \times 3 \times 3 \times 3$ which is the product of the sizes of all the variables' domains. In general this is always true: the number of candidate solutions is the product of the variables' domains' sizes.

Knapsack Problem

In this example we will model an instance of the decision variant of the 0–1 knapsack problem. Because that is a mouthful, we will henceforth just call it a knapsack problem!

The problem specification in ordinary language is this. Given a set of integers $\{7, 10, 23, 13, 4, 16\}$, a count 4, and a total 42, find a subset with 4 elements which sum to 42.

This problem can be modelled with six variables, one variable for every element of the original set. We could name the variables A, B, C, D, E, and F. Each variable has domain $\{0, 1\}$ meaning each variable can take one of two alternative values, either 1 (signifying that it is included in the subset) or 0 (signifying that it is not included).

There are two constraints. Firstly we require the included subset to have exactly four elements. This can be expressed in terms of the variables as a sum:
$A + B + C + D + E + F = 4$. Any assignment for the above variables that satisfies this constraint represents a set with four elements.

The second constraint is that the sum of the included elements is 42. This is expressed elegantly by the following equation:
$7 \times A + 10 \times B + 23 \times C + 13 \times D + 4 \times E + 16 \times F = 42$.

If, for example, A, C and F are included, then they are all assigned the value 1, while the remaining variables are assigned the value 0. This is the candidate solution $A = 1$, $B = 0$, $C = 1$, $D = 0$, $E = 0$, and $F = 1$. For this candidate, the previous equation then simplifies to $7 + 23 + 16 = 42$ which is false, so the constraint is violated, and this candidate turns out not to be feasible.

The Assignment Problem

The final problem modelled in this section is called the assignment problem. Wikipedia gives the following definition of the problem:

> There are a number of agents and a number of tasks. Any agent can be assigned to perform any task, achieving some value that may vary depending on the agent-task assignment. It is required to perform all tasks by assigning exactly one agent to each task in such a way that the total value of the assignment is maximized.

Typically the value of assigning each agent to each task is given as a table: The decision version of this problem asks if there is an assignment with value of at least a given limit. In this example the limit is 55 (Fig. 2.4).

We can model this problem with a decision variable for each agent, PA, PB, PC, PD with same domain for each variable: $\{task1, task2, task3, task4\}$.

The constraints impose that each agent does a different task:

$PA \neq PB, PA \neq PC, PA \neq PD, PB \neq PC, PB \neq PD, PC \neq PD$

Note that $X \neq Y$ ("X not equal to Y") holds if, and only if, X and Y take different values. The $\neq$ constraint can be imposed on two variables, or a value and a variable—or uselessly on two distinct values.

A way to model the value is to introduce another set of variables
$ValueA, ValueB, ValueC, ValueD$
to represent the value achieved by assigning each agent.

The constraint on the value variables are in Table 2.1.

This is rather long-winded and we will explore a much more succinct way to model cost in the next subsection.

The final constraint is $ValueA + ValueB + ValueC + ValueD \geq 55$.

Value of assignment	Job 1	Job 2	Job 3	Job 4
Agent A	18	13	16	12
Agent B	20	15	19	10
Agent C	25	19	18	15
Agent D	16	9	12	8

Fig. 2.4 The assignment problem

Table 2.1 Assignment values

If	PA	=	task1	then	ValueA	–	18
			task2				13
			task3				16
			task4				12
If	PB	=	task1	then	ValueB	=	20
	etc.						

The reader may have noticed that this same problem could be modelled another way, by introducing a decision variable for each task, *A1,A2,A3,A4* each with domain: {*agentA*, *agentB*, *agentC*, *agentD*}. In this case we would introduce a value variable for each task: *Value1,Value2,Value3,Value4*.

Many problems can be modelled in different ways, and sometimes the choice of model has a very strong effect on the computational effort needed to solve it.

Travelling Salesman Problem

The travelling salesman problem or TSP is perhaps the most famous of the *combinatorial optimisation* problems that are discussed in this text.

The problem first became well-known in 1954 when it was used for a competition in the USA. The problem is to visit a number of locations minimising the total travel distance. In our version, it is a travelling ecologist who must visit a number of cities worldwide to tell people about global warming. In this case it is the emitted CO_2 that must be minimized, but travel distance is a good proxy for CO_2 emissions!

The input data for the problem is a list of distances between the locations, and the only constraint is that the ecologist starts and ends at the same location. Assuming the distances satisfy the "triangle inequality"—the direct route from *A* to *B* is always shorter than going via a third location *C*—then the shortest route will visit each location exactly once (Fig. 2.5).

In the following we shall focus on the *decision variant* of the TSP (as we did for the previous problems). In this variant we seek a tour which covers all locations with a distance less than some give limit.

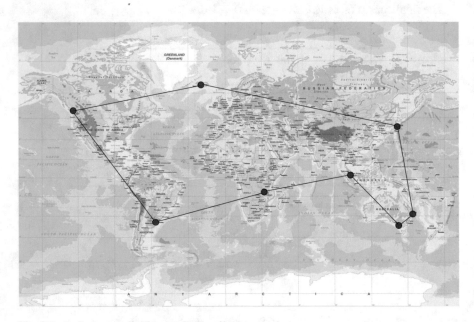

Fig. 2.5 An 8-city tour for the travelling ecologist

Table 2.2 Distances between cities

	ca	cb	cc	cd
ca	0	1000	2500	500
cb	1000	0	1500	3000
cc	2500	1500	0	2000
cd	500	3000	2000	0

To keep things short we start with an example TSP with just 4 cities: ca, cb, cc, cd. The distances between them are shown in Table 2.2. We must find a tour with distance of at most 60000.

Given that the ecologist must start and end at the same location, the total distance is the same whichever location she starts at. Assume, therefore she starts at ca. We introduce three variables $City2, City3, City4$, one for the city she visits next, after ca, one for the third city, and one for the fourth city.

We also assume she visits each city only once, so the domain of the variables $City2, City3, City4$ is $\{cb, cc, cd\}$.

There are also resulting constraints:
$City2 \neq City3$, and $City2 \neq City4$ and $City3 \neq City4$.

The length of the tour, given the choice of $City2, City3$ and $City4$ is:
if $City2 = cb, City3 = cc, City4 = cd$ then 5000
if $City2 = cb, City3 = cd, City4 = cc$ then 8500
if $City2 = cc, City3 = cb, City4 = cd$ then 7500
if $City2 = cc, City3 = cd, City4 = cb$ then 8500
if $City2 = cd, City3 = cb, City4 = cc$ then 7500
if $City2 = cd, City3 = cc, City4 = cb$ then 5000
Thus the problem is satisfiable with a solution $ca \rightarrow cb \rightarrow cc \rightarrow cd \rightarrow ca$ for example.

An alternative model introduces a variable for the successor of each city on the tour. We will use the name Sa for the variable whose value is the successor of a, Sb for the successor of b, and so on. Sa, Sb, Sc, Sd

The domains of these variables are:
$Sa : \{cb, cc, cd\}$
$Sb : \{ca, cc, cd\}$
$Sc : \{ca, cb, cd\}$
$Sd : \{ca, cb, cc\}$

This model makes it easier to specify cost variables: $Costa$ for the cost of travelling from ca to Sa, $Costb, Costc$ and $Costd$ similarly. Thus we can specify $Costa$ as:
if $Sa = cb$ then 1000
if $Sa = cc$ then 2500
if $Sa = cd$ then 500
etc.

The final constraint is $Costa + Costb + Costc + Costd \leq 6000$.

In this alternative model, to ensure the solution is a tour, surprisingly it is *not* sufficient to impose the constraints
$Sa \neq Sb$, $Sa \neq Sc$, $Sa \neq Sd$, $Sb \neq Sc$, $Sb \neq Sd$, and $Sc \neq Sd$.
We leave it as an exercise for the reader to determine why these constraints are not sufficient to ensure that any feasible assignment to the variables represents a single tour of all the cities (answer in next section!).

2.2.4 Modelling with Arrays

We can group a list of parameters into an "array", and we can also group a list of decision variables into a "variable array". For example the set of integers in the knapsack problem $7, 10, 23, 13, 4, 16$ can be represented as the array $[7, 10, 23, 13, 4, 16]$. The array does not seem much different from the original list of parameters! However the difference arises because

1. we can give a name to an array, e.g. $items = [7, 10, 23, 13, 4, 16]$ gives the name *items* to the above array
2. we can use an "index" to select any element of the array. For example $items[3]$ selects the third element of the array *items*. In this example $Item[3]$ denotes the value 23.
3. We can use a shorthand for various operations on arrays. The length of the array can be written thus $length(items)$ (which is 6). There is also a shorthand for the sum of the elements of the array. Mathematicians represent a sum using the Greek letter $\sum$ for capital "S" so $\sum items = 7 + 10 + 23 + 13 + 4 + 16 = 73$
4. We can combine the shorthands with array indices. So for example we can sum the first four elements of an array thus:
   $\sum_{i \in 1..4} items = 7 + 10 + 23 + 13 = 53$

To demonstrate the power of the array notation, let us revisit the knapsack, assignment and travelling salesman problems.

Knapsack Problem
The problem specification in ordinary language is this. Given a set of integers 7, 10, 23, 13, 4, 16, a count 4, and a total 42, find a subset with four elements which sum to 42 (as in Fig. 2.1 above).

This problem instance can be expressed neatly by putting the numbers and the variables into arrays.

We give the name *items* to the array $[7, 10, 23, 13, 4, 16]$, and the name *Vars* to the variable array $[A, B, C, D, E, F]$, and then $Vars[1]$ is the variable A. Remember that for the "indicator" variables in the array A, B, C, D, E, F, they take the value 1 if the corresponding item is included in the knapsack. 0 indicates that it is left out. Thus in the array *Vars*, $Vars[i]$ takes the value 1 if and only if the ith item is included in the knapsack. For example, the assignment $A = 1, B = 0, C = 1, D = 0, E = 0$, and $F = 1$, can simply be expressed as $Vars = [1, 0, 1, 0, 0, 1]$. The number of

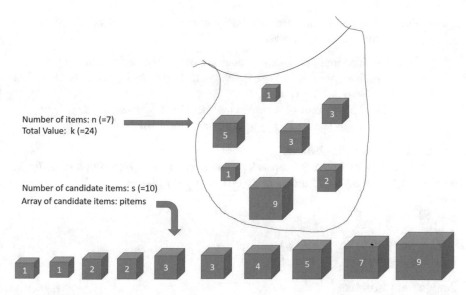

Fig. 2.6 Knapsack data

items in the knapsack is specified as a parameter value 4, and the target total value as a parameter value 42.

The constraint $A + B + C + D + E + F = 4$ can now be expressed in the following notation $\sum Vars = 4$.
The constraint $7 \times A + 10 \times B + 23 \times C + 13 \times D + 4 \times E + 16 \times F = 42$ is expressed as $\sum_{j=1..6}(items[j] \times Vars[j]) = 42$.

Instead of modelling a single instance of the knapsack problem, we can now model the knapsack problem class. An instance of the knapsack problem is specified by giving each parameter a value.

The number of items in the knapsack is the parameter n and the target total value is k. Let us use new names for the array of parameters *pitems* and for the variable array *PVars*. We make the length of the arrays a new parameter s. These are illustrated in Fig. 2.6.

Parameters   There are three integer parameters

- The count of included numbers n
- The sum of the included numbers k
- The length of the (input) set of candidate items s

The other parameter is the array *pitems* of candidate items (of the right length S)

Variables   To model this problem class we introduce an array *PVars* of n variables, each having the domain $\{0, 1\}$ as before

Constraints   The constraints are then easily expressed in terms of n, k,s, *pitems* and *PVars* as follows:

- The input parameter values are consistent: $length(pitems) = s$
- The variable array has the correct number of decision variables: $length(PVars) = s$
- The number of included numbers is n: $\sum PVars = n$.
- Their sum is k: $\sum_{j=1..s}(pitems[j] \times PVars[j]) = k$.

The model is thus quite abstract.

We can define an arbitrary knapsack problem by simply specifying a value for n, a value for k, a value for s, and an array of s values for *pitems*:

$n = 5$
$k = 58$
$s = 8$
$pitems = [5, 7, 9, 10, 14, 16, 17, 18]$
(A solution is $PVars = [1, 0, 1, 1, 0, 1, 0, 1]$.)

Assignment Problem

In the assignment problem there are a number of agents and a number of tasks. Any agent can be assigned to perform any task, achieving some value that may vary depending on the agent-task assignment. It is required to perform all tasks by assigning exactly one agent to each task in such a way that the total value of the assignment is maximized.

We can either model this problem with a variable for each agent, whose value represents the task they perform, or else we can model the problem with a variable for each task, whose value is the agent who performs that task. Arbitrarily we here present the model with a variable for each task.

To model the assignment problem we will use a two-dimensional array, which is essentially an array of arrays, or a *matrix*. A matrix can be used to represent a table of parameters—in this case the value achieved by assigning different tasks to each agent. The agents, $p1$, $p2$, $p3$, $p4$ might be carpenters with different levels of skills and experience; and the tasks $j1$, $j2$, $j3$, $j4$ might be roofing jobs on different houses under construction. For our example problem the values can be represented by the 4×4 matrix in Table 2.3.

A matrix allows us to use two indices to select an element. The first index selects the row and the second index selects a column. Thus in the above example $values[p2, j3] = 19$.

Table 2.3 Value of assigning each person to each task

		Tasks			
		j1	j2	j3	j4
Agents	p1	18	13	16	12
	p2	20	15	19	10
	p3	25	19	18	15
	p4	16	9	12	8

We will also introduce a variable array $A = [A1, A2, A3, A4]$ to represent the agent assigned to each task. Recall that the syntax $A[1]$ denotes the first variable $A1$, $A[2]$ denotes $A2$ and so on.

We introduce some new shorthand when presenting these constraints. An example of this shorthand is to express the three constraints
$A[1] \neq A[2]$, $A[1] \neq A[3]$, $A[1] \neq A[4]$ as a single constraint:
$A[1] \neq A[j] : j \in 2..4$
We can read this as "$A[1] \neq A[j]$ for each j in the set $2..4$"

Indeed if we want to ensure $A[i] \neq A[j]$ for every pair i and j in $1..4$, where i is not the same as j, we can write
$A[i] \neq A[j] : i \in 1..4, j \in 1..4$ where $i \neq j$

Alternatively, because $A[i] \neq A[j]$ is the same constraint as $A[j] \neq A[i]$, we can write:
$A[i] \neq A[j] : i \in 1..3, j \in (i + 1)..4$

Parameters	– The number of agents, and tasks, n
	– An $n \times n$ matrix of values, *value*. The cost of assigning agent i to task j is $value[i, j]$
	– A limit *limit* on the total cost of all the assignments.
Variables	– An array A of n variables, where the value of the variable $A[j]$ represents the agent performing the jth task. The domain of each variable is $1 \ldots n$, representing the possible agents who can perform that task.
	– An array TV of n variables (TV standing for Task-Value), where $TV[j]$ is the value of performing the jth task. This value is inferred from the choice of agent to perform the task. The domain of the variables in TV is the set of values that appear in the matrix *value*.
	In case a particular agent cannot do a certain task (because they lack the necessary skills, perhaps), then the domain associated with that task variable simply excludes the value representing that agent.
Constraints	– The values of all the variables in the array A are different: $A[i] \neq A[j] : i \in 1..(n - 1), j \in (i + 1)..n$
	– The task-value of the jth task is the value of assigning the chosen agent to that task: for each $j \in 1 \ldots n, TV[j] = value[A[j], j]$.
	– The total value is at least the given limit *limit*: $\sum TV \geq limit$.

An instance of the assignment problem is specified by giving values to all its parameters: $n = 3$, the matrix *value* is given in Table 2.4, and $limit = 14$.

This is satisfiable with: $A = [3, 2, 1]$ because agent $A[1]$ (i.e. 3) performs task 1, with a value given as $value[3, 1] = 7$; agent $A[2]$ (i.e. 2) performs task 2 with $value[2, 2] = 4$; and agent $A[3]$ (i.e. 1) performs task 3 with $value[1, 3] = 3$. This gives a total value of $7 + 4 + 3 = 14$.

Table 2.4 Assignment matrix

		Tasks		
value =		1	2	3
Agents	1	1	2	3
	2	4	4	4
	3	7	6	5

Travelling Salesman Problem

The travelling salesman problem (or travelling ecologist problem as illustrated in Fig. 2.5) is to find the shortest circuit starting at one location, passing through a given set of other locations, and returning to the start.

We can model the travelling salesman problem class in two ways which superficially appear similar.

TSP—First Model

Parameters – Number of cities n. The cities will be represented in the model by the numbers $c1..cn$.
  – Distance between cities, represented as an $n \times n$ matrix *distance* where *distances[ci,cj]* is the distance between cities ci and the cj.
  – A limit *limit* on the total distance of the tour.
  Each leg of the tour will start at one city ci and end at another city cj, thereby incurring the travel distance *distance[ci,cj]*.

Variables – A variable array *Vars* of length n, one variable for each city. The value of the variable *Vars[j]* represents the city visited jth in the tour. Each variable has domain $c1 \ldots cn$.
  – A variable array *Cost* of n travel distance variables one for each leg of the tour. The domain of each travel cost variable is the set of all distances (i.e. costs) in the distance matrix.
  For example if $n = 4$, the assignment *Vars* $= [c3, c1, c4, c2]$ represents the tour in which city c3 is visited first (because it is assigned position 1 in the tour), city c1 is second, city c4 is third and city c2 is fourth. This assignment denotes the complete tour $c3 \rightarrow c1 \rightarrow c4 \rightarrow c2 \rightarrow c3$. This represents a tour with four stops, starting and ending at the first stop c3.
  The value of the jth variable *Cost[j]* in the array represents the travel distance of the jth leg of the tour. Specifically, the travel distance of the jth leg of the tour is the distance from the city in position j in the tour, to the city in position $j + 1$.

Constraints – The tour should visit all n different cities—i.e. all the variables in *Vars* must take distinct values:
  $Vars[i] \neq Vars[j] : i \in 1..n, j \in (i + 1)..n$.

- The travel distance of the jth leg of the tour is the distance from the city in position j in the tour, to the city in position $j + 1$: if $X = Vars[j]$ and $Y = Vars[j + 1]$, then $Cost[j] = distance[X, Y]$.
  This can be expressed as the constraint:
  $(Cost[j] = distance[Vars[j], Vars[j + 1]]) : j \in 1..(n - 1)$
  and a second constraint to deal with the nth constraint:
  $Cost[n] = distance[Vars[n], Vars[1]]$
- The final constraint is that the total travel distance of the tour is below the given limit *limit*:
  $\sum Cost \leq limit$.

TSP—Second Model

The second model of the TSP has the same distance matrix. Similarly there are n city variables and n travel distance variables. We replace the variable array *Vars* with a new variable array V.

The array V has indices which are cities, rather than numbers:
$V[ci] : ci \in c1, \ldots cn$
The element $V[ci]$ no longer represents the city visited ith in the tour, but instead it represents the city visited after city ci. In this case the assignment $V = [c3, c1, c4, c2]$ now represents the tour $c1 \rightarrow c3 \rightarrow c4 \rightarrow c2 \rightarrow c1$.
The value of the jth travel distance variable is now the distance to the next city after city cj. $Cost[cj]$ therefore represents the travel distance from cj to $V[cj]$—the city visited after city cj.

The two representations are contrasted in Fig. 2.7.

The advantage of this model is that the constraint relating the city variables to the travel distance variables is simpler to evaluate since it only depends on the value of one decision variable:
$(Cost[cj] = distance[cj, V[cj]]) : cj \in c1 \ldots cn$.

There is a disadvantage, and we encountered it above. Under this model the constraint that all the variables in V take different values, would not be enough to

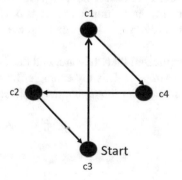

Model 1: Index: 1 2 3 4
Vars = [c3,c1,c4,c2]

Model 2: Index: c1 c2 c3 c4
V =   [c4,c3,c1,c2]

Fig. 2.7 Two alternative models for the TSP

guarantee that the tour visited all the cities. Consider, for example, the assignment $[c2, c1, c4, c3]$. This represents two unconnected tours: $c1 \rightarrow c2 \rightarrow c1$ and $c3 \rightarrow c4 \rightarrow c3$!

A constraint that ensures that V represents a single tour (or circuit) is quite awkward to capture.[5]

Perhaps surprisingly this second model is the one usually employed to represent the TSP.

2.3   Summary

This chapter has introduced *algorithms* and *models*, and how they are related. Models involve variables, constraints and parameters. A problem class is defined by a model, and a problem instance is specified by giving the values of the parameters. A solution is an assignment of values to the variables that satisfies the constraints.

Models for problem classes can be made more compact using arrays and matrices. They can be arrays (or matrices) of parameters or of variables. An index enables a specific element of an array or matrix to be selected, and sets of indices selects sets of such elements.

The chapter has developed models for a set of example problems, and has illustrated how the same problem can sometimes be captured by alternative models.

Answer to the Breaking Chocolate Problem   How many steps are needed to break the bar into 117 pieces?

The answer is in this line of the problem statement:

"At each subsequent step you take just one piece, and break it into two". Accordingly each step produces just one more piece, so the minimum number of steps is 116.

It doesn't matter what strategy you use, since every one is optimal and every one takes 116 steps!

Most industrial problems sound more complicated than they are, and the Eureka moment is when you see how to make things simpler! This toy example illustrates that you cannot develop (good) IDSS solutions without *intelligent developers*!

---

[5] In Minizinc it is expressed as a "global" constraint—see Sect. 8.4.

Chapter 3
Examples of Industrial Decisions

The core requirements addressed by the models introduced in Chap. 2, such as packing, assignment and routing, occur in many industrial applications. However they do not reflect the complexities imposed by practical requirements.

This chapter introduces a variety of substantial practical applications for which IDS has been developed. While it is instructive to describe these different applications, their modelling is beyond the scope of this text. Indeed writing out the model of the supply chain application described below required several thousand lines.

The applications are presented in this chapter under the themes of *strategic*, *tactical* and *operational*.

3.1 Strategic Optimisation Examples

3.1.1 Case Histories

Airline Network Planning
An airline plans its network of airports and interconnecting routes years in advance of the target flying season. The challenge is to decide between which pairs of airports there should be regular flights, and how many aircraft should fly on each route.

The data on which the decisions depend is termed "Origin-Destination" data. This is the data about how many people want to fly from each origin to each destination at each time of each day. The data has to be inferred from historical flight data. Although historically there may have been no (or few) direct connections from a given origin to a given destination, demand for this origin and destination can be inferred from historical multi-leg journeys from the original to one or more intermediate airports, before finally reaching the destination. Data must be collected from all airlines —and if possible from other forms of transport as well—in order to complete the picture.

© Springer Nature Switzerland AG 2020
M. Wallace, *Building Decision Support Systems*,
https://doi.org/10.1007/978-3-030-41732-1_3

The next aspect is to consider the type of network run by the airline. If, for example, the airline runs a "hub-and-spoke" network, with all aircraft maintenance and infrastructure located at a single airport—the hub—then changing to a multi-hub or spoke-to-spoke network may not be possible. Even if the type of network is fixed, however, there remain decisions to be made about which pairs of airports to serve with direct flights, which to serve via intermediate stops, and which not to serve at all [3].

Even given an origin-destination demand, the services required by different passengers can be quite different [4]. Business passengers are prepared to pay more to meet specific time constraints, while leisure passengers are more fussy about price, but less about time. Three type of passengers can thus be distinguished:

1. Business passengers travelling in the morning, who need to arrive at a specific time
2. Business passengers travelling in the evening, who wish to depart at a specific time
3. leisure passengers who seek the cheaper flights

Historical data can be mined to determine the proportion of each type of passenger with the given origin and destination. The willingness of passengers to fly via an intermediate airport, rather than taking a direct flight, depends on the extra time required, and the passenger type.

Based on all this data, the decision problem is how best to meet this demand with a given fleet of aircraft. It reduces to the question of what flights at what times should be scheduled on each route covered by the network. The decision problem is to find the set of flight-legs which can be performed by the available aircraft fleet, and maximise the value of passenger demand met by these flights. Naturally there is scope to expand the fleet to some extent.

Take-off and landing slots are a significant constraint at major airports: indeed airlines sometimes have to pay large sums of money to the previous holder of a slot that they need. Retiming flights to fit the available slots is a challenging optimisation problem.

Competitor airlines will also address the same passenger demand, and a key challenge is to understand how to deliver the best service, without yet knowing what service the competitor will offer. Accordingly the precise times for the flights are sometimes not decided during strategic planning: merely which (combinations of) flights can meet the demand of the different types of passengers. These flight times can be firmed up as better information about the demand and the competitors emerges.

Planning the Size of the Work Force

In industries involving special skills and training—for example health and the military—to reach a future level of skilled personnel, plans for training and experience must be established many years in advance. Any substantial change in personnel—such as opening or closing a department in a University—requires strategic planning. The next case involved an automobile rescue service, with a large fleet of rescue vehicles and patrollers. A new competitor had appeared with a different business model. This company had no fleet or engineers, but simply dispatched every call to a local garage. The question arose: is it best to have a fleet and employ your own patrols, or is it better to outsource everything to contractors?

The decision support to address this question can be tackled by using the historical data to calculate the operational costs of running the operation with a smaller or larger fleet. Taking a period of time, for example a year, a simulation of the year's operations could be constructed.

In this case the assumption must be made that the outsourcing will be managed well, so that the change will not impact the number of customers for the service. Thus historical data should be a good guide to future demand. (If there is a steady rate of increase, or decrease, that can be factored in.)

For each rescue request the marginal cost of sending a patrol is (effectively) zero, but each rescue by a contractor incurs as the marginal cost, the full cost of the contractor. The service fleet and patrollers incur an annual fixed cost instead. The optimisation model can then run a year's rescue requests, and optimise the proportion of patrols and contractors in order to minimize the total cost.

In the context of the simulation, every decision taken during a whole year by all the dispatchers in the company must be handled by the IDS system. Thus instead of decision-support, the simulation uses automated decision-making. A dispatch decision must take into account the location of the job, the location of each patrol, the time when it is likely to become available, and the time required to travel to the job location. The decision must also take into account the estimated time and cost to dispatch a contractor. The objective is to maintain a target response time for (almost) all calls and to minimise total cost.

Finally it is important to notice that specifying an IDS requirement is never easy. The discussion up to this point has only considered financial implications of reducing the service fleet. This is only an approximation of the real problem. There is a drawback to using contractors. On a cold and rainy day when there are many breakdowns, all the garages become very busy and either contractors are unavailable or they can charge very high prices. In this case the company needs its own patrols to guarantee a service and avoid the response times becoming unacceptably high. This, and further factors not mentioned above, come into play in major strategic decisions of this kind.

3.1.2 Issues in Strategic Planning

Intelligent decision support, in the form of optimisation, has a role in strategic decision-making, but realistically only a supporting role. At strategic timescales there are relevant factors that cannot be predicted from historical data. The results from the decision support system must be augmented by expert judgement.

3.2 Tactical Decision Support

Tactical decision-making is required at a point where the resources have been put in place and the question is how best to deploy them.

3.2.1 Case Histories

Production Planning

The deployment of personnel and resources to meet demand arises both in manufacturing and the service industry. A fortune 500 company was producing packaging at facilities around the world. Two aspects of the production were the construction of specific containers, and the printing of information on the containers.

The containers were constructed on various machines using ancillary equipment which needed to be loaded for different types of container. Consequently each container needed to be constructed on the correct machine type, loaded with the correct ancillary equipment. Changing the equipment on a machine takes time and human resources.

The printing was performed by printers which could use a fixed number of colours at any one time. Changing the colours on a printer was another process that took time and human effort.

The sequencing of products on the different machines dictated the number of changes—of colours on the printers and ancillary equipment on the packaging machines. Optimising the sequence minimised the personnel required and maximised the throughput [50]. However the optimal sequence for printing and the optimal sequence for container construction are typically quite different.

In this example one outcome of optimisation was unexpected: the optimisation revealed that the limited quantity of ancillary equipment was a bottleneck on productivity.

Supply Chain Optimisation

For production planning the decisions are the choice of which machines are used for each product, and the order in which the products are processed on the machines. Supply chain optimisation involves a much broader set of decisions. An Australian company imported some 15 varieties of a bulk commodity in order to meet a fortnightly demand for these varieties at multiple locations around the country. The decisions to be coherently optimised were:

- how much of each variety to order from each of a number of suppliers based in different countries in different continents: a large order across the whole season benefited from a reduced price for certain suppliers and products.
- How many ships of different sizes to use each fortnight to transport the commodities to Australia
- The configuration of the loads on each ship (how much of which commodities in which holds)

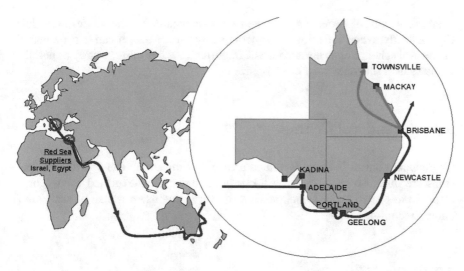

Fig. 3.1 The Supply Chain Application

- the route taken by each ship potentially visiting multiple suppliers and multiple Australian ports
- the quantity of each commodity unloaded at each Australian port
- the amount of each commodity held in storage at the port and the amount transferred by truck to the end customer

The application is illustrated in Figure 3.1.
All these decisions were constrained in different ways, for example:

- a ship which visited certain countries could not visit certain other countries on the same trip (for reasons dictated by the politics of the countries)
- a ship could only have one variety of commodity in each hold
- a fully laden ship could not enter certain Australian ports

The obstacle to true optimisation in this application was the impact of weather on fortnightly demand. The actual demand could vary greatly from the forecast demand in times of extreme weather (drought; floods etc.). Sadly Australia is the land of droughts and flooding rains! To cope with this uncertainty the storage levels in the warehouses in the Australian ports provide a buffer and the supply chains decisions were made with robustness to weather variations taken into account.

3.2.2   Issues in Tactical Scheduling

Tactical scheduling starts with a known set of resources, and the demand forecast in these timescales (weeks or months) is more reliable than when planning strategy. Consequently intelligent tactical decision support can deliver great benefits. Nevertheless on the day unexpected events frequently happen. Too often in these cases, operational control decisions are taken without regard for the tactical schedule. Consequently, when something goes wrong early in the day, it can happen that all the tactical scheduling is discarded. The importance of returning as early as possible to the tactical schedule must be recognized.

3.3 Operational Control

Operational control is an undervalued area of optimisation. The operational decisions are ones which have to be taken in response to unexpected events, or in response to requirements which arise at the very time when a plan or schedule is under execution.

3.3.1 Case Histories

Workforce Management
A supplier of office equipment also provided maintenance of the equipment, with engineers responding to customer calls by going onsite to fix the problems. As calls came in the decision support system was required to allocate the jobs to the right engineers. Clearly the engineer assigned to a job needed to be trained in the maintenance of the relevant equipment, and to be geographically close enough to the customer that travel time was minimised. Also a larger travel time not only results in a longer response time for the customer, but also it absorbs the engineer's time preventing her from responding to other calls. The time to complete a job can be short or long, and when a job takes a long time subsequent jobs which would have been assigned to the same engineer must be replanned.

A side-effect of optimisation in this case was a small change to the operating procedures of the engineers: when they arrived onsite and looked at the problem they reported back to the dispatch staff to give an estimate of the time required to complete the job. This simple change enabled improved planning that reduced the average response times and increased the number of jobs that could be completed by the workforce each day.

Rostering
Rostering for a period such as six weeks ahead could be classified as tactical. On the other hand rostering and re-rostering often continues right up to the start of the shift. A rostering problem typically comprises:

- a set of possible shifts (e.g. Day, Evening, Night, Off)
- a time horizon (e.g. six shifts)
- a set of employee types (skills; experience; seniority)
- a coverage requirement for each shift (the minimum number of staff of each type)
- a set of rules constraining which sequences of shifts are allowed (e.g. not a night followed by a day shift)

- a set of fairness requirements (e.g. all staff should work a similar number of weekends)
- a set of employees
- for each employee, their type and availability (typically shifts or days unavailable)
- for each employee, a set of preferences

Clearly rostering problems are highly constrained, and often when staff are ill, other staff may end up working two shifts in a row—violating the rules above. An intelligent rostering system can roster a team of employees to meet constraints and maximise fairness. For teams with over 20 employees rostering staff have great difficulty meeting the requirements, and constructing fair rosters.

The most awkward aspect of rostering is the handling of employee preferences. If some employees state multiple preferences in order to achieve an ideal roster for themselves, fairness suffers. Sometimes a preference reflects serious issues for the employee: for example a caring responsibility at home. In an organisation the rostering staff make judgements about which preferences to respect. These judgements are rarely supported by any records of an objective basis. The managers of these staff consequently have difficulty assessing the quality of a roster. The result is that, even if an intelligent automated rostering system produces much better rosters than the rostering staff, the organisation may be unable to recognize this. Consequently IDS for rostering is often rejected for the wrong reasons.

Passenger Rail Rescheduling

This example was not a typical operational control application, but illustrates a different way operational decision support can be deployed. The rail operators in this case had punctuality goals and penalties for failing to meet them.

The decision support software is run when a train is delayed, given an estimate of the period of the delay. The software then slows down some trains and skips stations on others, so as to minimise the impact of the delay on all the passengers in the rail network. The problem is complicated because a delay to one train also delays the following trains, and thus also causes delays on future trips assigned to those trains. Moreover when a train is full, many passengers waiting at the next station are unable to board; and when passengers are boarding the boarding time can become much longer than normal. The perceived delay and disruption for a passenger depends on several factors, such as how long the passenger is left standing on the platform, how crowded the train is and how much later she arrives than was scheduled [7].

An interesting aspect of this application is the requirement to get back onto the original schedule as soon as possible. It turned out that prioritising getting back on schedule over the objective of minimising total passenger delay had little effect on the passenger delay but made the operational control problem tractable. Indeed, getting back to the tactical schedule was a priority noted in the previous section: it avoids many operational side-effects, since when things are back to normal people and systems tend to function more efficiently.

While minimising total passenger delay and disruption was the objective, it emerged that this was quite different from minimising the penalty for the train operator. In short the train operators minimising their penalty resulted in unnecessary additional passenger delays and disruption.

3.3.2   Issues in Operational Control

The challenge in operational control is to find a solution quickly. In this case optimality is less important than finding a workable (feasible) solution. The algorithms used for operational control are therefore somewhat different from those used in strategic planning and tactical scheduling.

3.4   Intelligent Decision Support in Practice

The models in the previous chapter supported the optimisation of well-defined objectives on problem instances with known data. The examples in this chapter reveal some of the complexities that arise in real life: data based on uncertain forecasts, multiple objectives that are not always clear—and may indeed be changed in the course of designing the model.

A major issue is how to measure the quality, effectiveness and impact of an IDS system. Even if the decision support is correct, there may be factors that can only be taken into account by the human decision maker. An example from the car rescue strategy IDS, is the unpredictable customer-facing behaviour of contractors, as well as the potential for raised prices and longer response times for contractors on a wet day.

Another issue around decision support is lack of clarity about the objective to be optimised. For rostering, the perceived quality of a roster is different for an employee member who no longer receives all her preferences, than for the manager who sees proper staff coverage on all shifts. If the new rostering system violates too many preferences, staff may look for other jobs, with a strongly negative impact on the organisation.

Thus IDS is typically only used for decision support, usually in partnership with a human decision-maker.

Nevertheless IDS can play a major role in revealing the consequences of alternative decisions. Moreover the design and implementation of such a system can focus an organisation on key questions which need to be addressed in order to reach the correct decision. Finally IDS can reveal new business opportunities. An example in airline schedule construction was an intelligent solution for increasing aircraft utilisation. It turned out that the solution was directly applicable for negotiating aircraft landing and take-off slots, and could further be used to address flight leg punctuality.

3.5   Summary

This chapter has outlined a variety of industrial applications that have benefitted from intelligent decision support. Airline network planning, and car rescue work-force planning are two strategic applications. Tactical planning and scheduling have been applied in production optimisation and supply-chain optimisation and risk management. More immediate applications of decision support include engineer dispatch to maintenance tasks in geographically dispersed sites, rostering for shiftworkers, and transport rescheduling after a disruption. The applications reveal the key role of the human decision-maker in a decision support system.

Chapter 4
Problem Modelling in MiniZinc

This chapter introduces MiniZinc, a modelling formalism for expressing models such as those in the previous chapter. Formalising the model in MiniZinc, enables you to feed it to a computer, run it and get solutions. There are other modelling languages which enable you to run your model on a computer. Some languages, such as OPL [31] and Mosel [19], are designed as front ends to specific optimisation packages. Others, like AMPL [20], GAMS [21] and AIMMS [2] are designed to map to a class of optimization engines (specifically mathematical programming packages [67]).

The MiniZinc modelling language is used for demonstrations in this book because it is simple, powerful and frcc to usc for academic or commercial purposes. The download includes a variety of optimisation engines enabling MiniZinc models to be developed and run through the same IDE (user interface). MiniZinc can be downloaded from the website

```
www.minizinc.org
```

where there is also a tutorial, user guide and reference manual.

This chapter covers just enough MiniZinc syntax to enable the reader to formulate our example problems in MiniZinc. The MiniZinc tutorial [61] provides a more thorough introduction. MiniZinc is also presented in several online courses including Basic Modeling for Discrete Optimization [60] and Advanced Modeling for Discrete Optimization [59].

© Springer Nature Switzerland AG 2020
M. Wallace, *Building Decision Support Systems*,
https://doi.org/10.1007/978-3-030-41732-1_4

4.1   Variables, Constraints and Parameters

A model declares a set of variables, representing a set of decisions. Each variable
has a domain (its set of possible values): the optimisation engine will choose which
value when the model is run.

The constraints make the decision-making difficult, by ruling out incompatible
choices. Each constraint involves a set of variables (the variables in its scope), and
rules out combinations of values for those variables.

Thirdly the parameters specify an instance of the problem. The parameter values
give both its size and its input data.

4.1.1   A Model for a Simple Problem

We start with the simple boat problem, introduced, on page 15 above. There are five
variables, *Boat*1 ... *Boat*5 which will represent the decisions about which dock
each boat is allocated to. There are three docks *dock*1, *dock*2, *dock*3.

In MiniZinc these will be enumerated as follows:
```
enum docks = [dock1, dock2, dock3] ;
```
Now the set of values for each variable is the set docks

In MiniZinc we declare a variable (*Boat*1 for example) thus:
```
var Boat1 : docks ;
```
giving its domain.

The constraints in this problem prevent *Boat*1 and *Boat*2 taking the value
*dock*1. The first constraint is written:
```
constraint Boat1 != dock1 ;
```
where   != means "not equal to".

Finally the model specifies whether there is something to optimise, or whether
just to seek any choices that satisfy the constraints. For the boat problem we just
need the latter, so the MiniZinc syntax is:
```
solve satisfy ;
```
If MiniZinc is downloaded the MiniZinc IDE can be invoked to accept an input
model. The model in the MiniZinc IDE is shown in Fig. 4.1.

This model can be run by pressing the run symbol, and the optimisation engine
writes the solution to the output panel below.

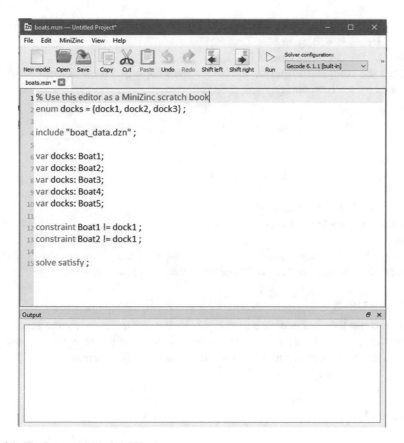

Fig. 4.1   The Boat model in MiniZinc

4.2   Modelling the Knapsack Problem

4.2.1   Arrays

In this section the knapsack problem, introduced, on page 16 above, will be
modelled in MiniZinc using arrays. The problem is to select four members of the
array [7, 10, 23, 13, 4, 16] that add up to 42.

In MiniZinc an input array has to be declared with its set of indices and the type
of its elements. Our array has indices 1..6 and its elements are integers. Accordingly,
the input array is specified in the MiniZinc model as

```
array [1..6] of int: items = [7,10,23,13,4,16] ;
```
The next item in the model is an array of variables with domain 0, 1. In MiniZinc
this is:
```
array [1..6] of var 0..1 : Vars ;
```

The syntax for a sum in MiniZinc uses the array indices, so the expression $\sum_{i \in 1..4} items$ is written in MiniZinc as:

```
sum(i in 1..4)(items[i])
```

Thus the complete MiniZinc model for the knapsack problem, with comments preceded by a % (percentage sign), is as follows:

```
% Input numbers
array [1..6] of int: items = [7,10,23,13,4,16] ;

% Zero-one variables - meaning included or not
array [1..6] of var 0..1: Vars ;

% Exactly four numbers should be included
constraint sum(i in 1..6)(Vars[i]) = 4 ;

% The total of the included numbers should be 42
constraint sum(i in 1..6)(Vars[i]*items[i]) = 42 ;

solve satisfy ;
```

There are not four items whose sum is 42, so the optimisation engine reports that the model is unsatisfiable. It is more fun to replace 42 in the above model with 43, and see the optimisation engine find the set of items which sum to 43.

4.2.2   Data

The knapsack model can be made more general by putting the parameter values in a separate file from the model. In this way the same MiniZinc model can be used for all problem instances in the knapsack problem class.

Suppose the parameter values are put in a file called values.dzn. The number of items to be included and the target total can also be input as parameters, and the model is as follows:

```
% Parameter specification
int: included ;
int: target ;
int: itemCount ;
array [1..itemCount] of int: items ;

% Load the file of parameter values
include "values.dzn" ;

% Model
array [1..itemCount] of var 0..1: Vars ;
constraint sum(i in 1..itemCount)(Vars[i]) = included ;
```

```
constraint sum(i in 1..itemCount)(Vars[i]*items[i]) = target ;
solve satisfy ;
```

The file `values.dzn` comprises the following parameter values:

```
included = 4 ;
target = 42 ;
itemCount = 6 ;
items = [7,10,23,13,4,16]
```

4.2.3 Knapsack with Item Weights and Values

With this syntax it is easy to model a more typical version of the knapsack problem where each item has a weight and a value. The knapsack has a limited capacity (which limits the total weight that it—or the person—can carry). The challenge is to maximise the value of items carried in the knapsack. This time the number of items carried is not limited—only their total weight.

Here is the model:

```
% Parameter specification
int: weightLimit ;
int: itemCount ;
array [1..itemCount] of int: weights ;
array [1..itemCount] of int: values ;

% Load the file of parameter values
include "values2.dzn" ;

% Model
array [1..itemCount] of var 0..1: Vars ;
var int: TotalValue ;

constraint sum(i in 1..itemCount)(Vars[i]*weights[i])
    <= weightLimit ;
constraint sum(i in 1..itemCount)(Vars[i]*values[i])
    = TotalValue ;

solve satisfy ;
```

This model will find the total value of any set of items with total weight within the limit included in the knapsack. To find the set of items which have maximum total value it is only necessary to replace the line
```
solve satisfy ;
```
with the line
```
solve maximize TotalValue ;
```

4.3 Assignment and Travelling Ecologist

4.3.1 Constraining All the Items in an Array with "forall"

In case there is an array of variables [$X1$, $X2$, $X3$, $X4$] and we need to impose a constraint on all of them we can use the "and" operator which is written /\ in MiniZinc. Thus the constraint that all of the X's take a value different from 1, might occur in a model as follows:

```
array [1..4] of var int: X ;
constraint (X[1] != 1 /\ X[2] != 1 /\ X[3] != 1 /\ X[4] != 1) ;
```

In ordinary language we would say something much shorter such as "All the X's must be different from 1". In MiniZinc we can do the same thing—with a syntax similar to the syntax for sum that we saw earlier. The above model extract can be written in MiniZinc as follows:

```
array [1..4] of var int: X ;
constraint forall(i in 1..4)(X[i] != 1) ;
```

We can use more than one index in a *forall* expression. For example to say that the first two X's take different values from the last two we can write:
```
forall(i in 1..2, j in 3..4)(X[i] != X[j])
```
Often we would like to impose that *all* the X's take different values. However an $X[i]$ cannot take a different value from itself. MiniZinc enables us to put a condition on the indices—in this case that we require the index i to be different from the index j:
```
forall(i,j in 1..4 where i != j)(X[i] != X[j])
```
Finally, before leaving the topic of logical expressions, note that Chap. 8 will introduce other logical operators such as:

```
/\    % and
\/    % or
not   % not
->    % implies
```

4.3.2 Assignment Problem

The small assignment problem (Sect. 2.2.4) and the travelling ecologist (Sect. 2.2.4) are both best modelled using two-dimensional arrays, also called "matrices".

In MiniZinc we start by listing the people and the tasks, as follows:

```
enum people = {p1,p2,p3} ;
enum tasks =   {t1,t2,t3} ;
```

The value from assigning each person to each possible task is held as a matrix (called "value") which can be declared as follows:

```
array [people,tasks] of int: value =
  [| 1, 2, 3 |
     4, 4, 4 |
     7, 6, 5 |] ;
```

The model includes the declaration of the parameters, but the parameter values are normally included in a separate data file. The following model returns an assignment of people to tasks whose value exceeds a limit of 58:

```
enum people ;
enum tasks ;
array [people,tasks] of int: value ;
set of int: vals;
int: limit ;


include "assignment_data.dzn" ;


array [tasks] of var people: A ;
array [tasks] of var vals: TV ;
var int: TotalValue ;


constraint
    forall(t1 in tasks, t2 in tasks where t1 != t2)
          (A[t1] != A[t2]) ;
constraint
    forall(t in tasks)(TV[t] = value[A[t],t]) ;
constraint TotalValue = sum(t in tasks)(TV[t]) ;
constraint TotalValue >= limit ;


solve satisfy ;
```

vals is the set of values in the matrix value.

The data held in the "assignment_data.dzn" file is this:

```
people = {p1,p2,p3} ;
tasks = {t1,t2,t3} ;
limit = 11 ;
value =
  [| 1, 2, 3 |
     4, 4, 4 |
     7, 6, 5 |] ;
vals = {1,2,3,4,5,6,7} ;
```

(In Sect. 8.2, below, we will see how to create such a list automatically from such a matrix).

4.3.3   *Travelling Ecologist Problem*

Section 2.2.4 above, presents a model of the travelling salesman problem. In this
section we formalise it in MiniZinc.

```
% Parameters

% Number of locations
int : n ;
% Distance between locations
array [1..n,1..n] of int : distances ;
% A limit on the total distance of the tour.
int : limit ;

% Assume the data is in a file called "ecologist-data.dzn":
include "ecologist-data.dzn" ;

% For the domain of travel distances,
% compute the maximum distance in the matrix distances
int : maxDist = max(i,j in 1..n)(distances[i,j]) ;

% Variables

% An array of n variables, one for each location, called V
% In this model the value of the variable  V[j]
% represents the location visited jth in the tour.
array [1..n] of var 1..n : V ;

% An array of N travel distance variables one for each leg of
  the tour.
array [1..n] of var 1..maxDist : TD ;

% The tour should visit all n different locations, which
  is equivalent
% to saying all the variables in the array V  must take
  distinct values.
constraint forall(i in 1..n,j in i+1..n)(V[i] != V[j]) ;

% The travel distance TD[j] of leg j of the tour is the distance
  % from the location in position j in the tour V[j], to V[j+1]
constraint forall (j in 1..n-1) (TD[j] = distances[V[j],V[j+1]]) ;
% and the final leg returns to the first location
constraint TD[n] = distances[V[n],V[1]] ;

%The total travel distance of the tour must be below the
  given limit
```

```
constraint sum (j in 1..N) (TD[j]) <= limit ;

solve satisfy ;
```

The file "ecologist-data.dzn" could have the following data, for example:

```
n = 4 ;
distances =
    [| 0,1,2,3
     | 4,0,3,5
     | 3,7,0,5
     | 4,3,2,0 |] ;
limit = 11 ;
```

Notice that according to these parameters, the distance from location 1 to location 2 (*distances[1,2]=1*) is different from the distance from location 2 to location 1 (*distances[2,1]=4*). When the distance in both directions is the same we call it a "symmetric" TSP. Another feature of these parameters is that the distance from location 3 to location 2 (distances[3, 2] = 7) is greater than the distance from 3 to 1 (3) plus the distance from 1 to 2 (1). Typically TSP problems satisfy the "triangle inequality" which says that the direct distance between any pair of locations is less than or equal to the distance via any intermediate location.

TSP which are symmetric and satisfy the triangle inequality are somewhat easier to solve.

4.4 Summary

This chapter has introduced MiniZinc through some examples. MiniZinc allows a model to enumerate the set of values in a data type. It supports arrays and matrices, and it enables the data to be separated from the model of the problem class. The objective of a MiniZinc model may be just to *satisfy* the constraints, or to *maximize* or *minimize* an objective expression. Finally to express something about all the values in a range, MiniZinc offers the *forall* construct. These features of MiniZinc are illustrated with models for the boats and docks, knapsack, assignment, and travelling ecologist problems.

4.5 Exercises

4.5.1 Worker Task Assignment Exercise

There are 12 tasks (numbered 1..12), and 6 workers (numbered 1..6). The following array of 0..1 variables

```
array [1..6,1..12] of var 0..1:Assign ;
```

represents decisions about assigning workers to tasks. $Assign[w, t] = 1$ if worker w is assigned to task t, and otherwise $Assign[w, t] = 0$.

First Challenge
Write a model which ensures at least 2 workers are assigned to each task.

Second Challenge
Extend your model to ensure that no worker is assigned to more than 5 tasks.

Third Challenge
Certain pairs of workers cannot be assigned to the same task. In the following matrix

```
array [1..6,1..6] of 0..1: bad_pair =
    [|0,1,0,1,0,0
     |1,0,0,0,1,0
     |1,0,0,1,0,0
     |0,0,1,0,1,0
     |0,0,1,0,0,0
     |0,0,0,1,0,0 |] ;
```

$bad\_pair[w1, w2] = 1$ means that workers $w1$ and $w2$ cannot be assigned to the same task.

Extend your model of the second challenge to meet the bad-pair constraint.

4.5.2 Knapsack Exercise

You have a bag which can carry 20 kg. You have a set of things you want to bring with you, and their weights:

```
enum items = {book, jacket, washbag, computer, boots,
              alarmclock, anorak, food} ;
array [items] of int: weight = [2,4,3,8,7,1,2,6] ;
```

First Challenge
These items have a certain value to you:

```
array [items] of int: value = [6,10,8,25,22,4,5,20] ;
```

Pack the items which you can carry in your bag that bring the highest possible total value to you.

Second Challenge
The knapsack also has limited space capacity, and the total volume of items it can fit inside is 2000 cm^2. Each item has not only a weight but also a volume:

```
array [items] of int: volume =
       [250, 500, 300, 250, 650, 130, 150, 600] ;
```

Fid the best solution, as for the first challenge, but the total volume of the items in the knapsack cannot be greater than the capacity of the knapsack.

Third Challenge

Some things are worth more in combination, and some less. Here is the additional (or reduced) score you get for each pair:

```
array [items,items] of int: extra_value =
    [|  0,  0,  0,-5,  0,  0,  0,  0
     |  0,  0,  0,  0,  3,  0,-2,  0
     |  0,  0,  0,  0,  0,  0,  0,  0
     |-5,  0,  0,  0,  0,-2,  0,  0
     |  0,  3,  0,  0,  0,  0,  0,  0
     |  0,  0,  0,-2,  0,  0,  0,  0
     |  0,-2,  0,  0,  0,  0,  0,  0
     |  0,  0,  0,  0,  0,  0,  0,  0
    |] ;
```

If $extra\_value[i1, i2] = 3$ and if items $i1$ and $i2$ are both in your bag then the total value of your bag is increased by 3. Naturally if $extra\_value[i1, i2] = -2$ then it is decreased by 2.

Extend your model for the first challenge to maximize the total with the modified values.

Chapter 5
Algorithms and Complexity

Instead of problem models, this chapter focusses on algorithms for solving them. Most of the models presented in this book are hard to solve in a way that is surprising: the algorithms required to solve them grow *exponentially* with the size of the problem data. This chapter makes explicit what is meant by "exponential" growth, and how we can assess the computational cost of running algorithms. Finally the chapter will explore what classes of problems seem to need exponential algorithms to solve them.

5.1 Combinatorial Problems

5.1.1 Chess

We start this section with the fable about the inventor of chess. The emperor of his country was so delighted by the game he offered the inventor anything he wanted. "I only want to cover all the cells on the chessboard with rice", answered the inventor, "one grain on the first cell, 2 on the second, 4 on the third, with twice as many grains on each subsequent cell, until all the cells are covered". Of course the emperor was amazed that the inventor should ask for so little, but he agreed and had a sack of rice brought in from which his servants would count out the prize.

The rest is history: the cell at the end of the first row, had $2 \times 2 \times 2 \times 2 \times 2 \times 2 \times 2 \times 2 = 2^5 = 256$ grains, weighing about 1 gram.[1] By the end of the fourth row, the cell required 2^{32} grains, which is about 4 billion, weighing in at about 10 tonnes.

---

[1] 2 *squared* is written 2^2; 2 *cubed* is written 2^3; 2^5 is called 2 *to the power of* 5. For any number N, N squared is N^2, 2^N is 2 to *to the power of* N. The current chess example shows that 2^N is much larger than N^2 for values of N greater than 4, and the difference grows bigger for larger values of N.

© Springer Nature Switzerland AG 2020
M. Wallace, *Building Decision Support Systems*,
https://doi.org/10.1007/978-3-030-41732-1_5

If the emperor had had enough rice to cover the last cell on the board, it would have amounted to about 10 tonnes of rice for every person on earth!

5.1.2   The Number of Candidate Solutions for Larger Problem Instances

Boats and Docks

In Sect. 2.2.3 above, we counted the number of candidate solutions for an instance of the boats and docks problem with 3 docks and 5 boats. With a model with a variable for each boat, and where the first two variables had a domain of size 2, and the other three boat variables a domain of size 3, the number of candidate solutions was $2 \times 2 \times 3 \times 3 \times 3 = 108$. More generally if there are N variables, with domains of size $D_1, \ldots D_N$, then the number of solutions is $D_1 \times D_2 \times \ldots \times D_N$.

Suppose, to keep things simple, all the boats could fit in any of the docks, so the domain size for every boat variable was 3. In this case the number of candidate solutions would have been $3 \times 3 \times 3 \times 3 \times 3 = 3^5 = 243$. Suppose next that the harbour opened three additional docks, and the harbourmaster's enlarged problem instance was to allocate not five but ten boats amongst the six docks. The number of candidate solutions now (assuming all boats could fit into any dock) is 6^{10}. This number is 60466176 which is more than 60 million! By doubling the size of the harbour, the harbourmaster has ended up with a problem which is 250,000 times larger than his original one!

He could deal with it, as many people do when faced with this kind of growth, by dividing the harbour into two parts, with three docks each, and delegating half the boats, i.e. five boats, to each part. This subdivided problem now has $3^5 = 243$ candidate solutions, for each of the two subproblems, which is manageable.

However, probably without realising it, the harbourmaster has thereby eliminated $60466176 - (243 \times 2)$ candidate solutions from consideration. In fact the two subproblems together admit only 0.0008 percent of the candidate solutions. The chance of the best solution being amongst this tiny percentage is also tiny!

Covering a Dockside with Shipping Containers

A wharf on the waterfront is marked out with lines forming an $N \times N$ grid pattern of squares much like a giant chessboard. The port authority uses this grid pattern to locate containers which have been lifted off ships and stacked using a crane. A standard-size container exactly covers two adjacent squares.

To make the best of the available space, we wish to cover the wharf completely with containers. Since the number of squares on the wharf is odd if N is odd we can only completely cover the wharf if N is even (Fig. 5.1).

The number of solutions to this problem grows very quickly with the size of the wharf N. The consequence is that even with a fast computer the time taken to find

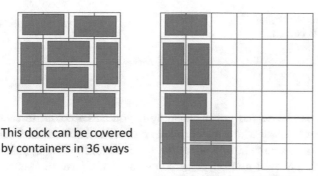

This dock can be covered
by containers in 36 ways

This can be covered 6728 ways

Fig. 5.1 Covering the wharf with containers

all the solutions grows quickly as well. To illustrate why this might matter—even if we do not actually need to find all the solutions—we introduce a variant of the problem.

In the variant, the port authority's chairman and CEO decide to go for a holiday in Bali. They park their company cars on the squares at the two opposite corners of the grid. The resulting grid is missing just two cells. In this variant we still have to cover the wharf with containers, but it has been modified by removing the squares from the two opposite corners of the grid.

As long as the number of rows and columns N is even, this modified wharf space still has an even number of squares. Can it be completely covered with containers if $N = 8$? and more generally for what values of N can it be completely covered? This is not in itself a real-life problem, but it does illustrate the kind of subproblem that comes up when real-life IDSS applications are under development.

The obvious approach is to use the same algorithm for this variant that was used for the original problem: keep adding containers until there are no more cells to cover. This is an inefficient and costly way to address this variant of the the problem. Try to think of a better approach.

It turns out quite easy to prove there is no way to cover the modified completely with containers. The naive computer program to prove this is to try all the different ways of placing the containers, failing each time when there are isolated cells left uncovered. The time taken to try all combinations and fail is almost the same as the time taken to find all solutions to the original problem of covering a normal wharf with containers. Indeed running the program described above yields the results in Table 5.1. These times naturally depend on the specific computer used to solve the problem, but the rate at which the time required increases with the size of the board is independent of the speed of the computer. With a board of size 12 or more the time is prohibitive on any computer!

Table 5.1 Time in seconds
to try and fail to cover a
modified wharf with
containers

Size of wharf	Time before failure
4	0.03
6	0.25
8	199.0

In fact it can be solved rather elegantly if you add a colour to each cell in the wharf—either black or white, and they are coloured alternately black and white like the squares on a chessboard. Notice that the cells in the two opposite corners of a wharf, with any even number of columns and rows, have the same colour (both black or both white). Since the original wharf, if N is even, has the same number of black and white cells, the modified wharf must have two less of one colour than the other (either two less black cells or two less white cells). You can also observe that any pair of neighbouring cells on the board have opposite colours. Consequently wherever you place a container on the wharf it must cover a black and a white cell. Therefore it is impossible to cover more cells of one colour than the other, however many containers you put on the wharf. It follows that the modified wharf can never be completely covered by containers. This holds for modified wharfs of any size N, where N is even.

5.2   Generate and Test

5.2.1   A Method for Testing a Finite Number of Candidates

In discussing the boats and docks problem, we worked out how many candidate solutions there were, based on the number of variables in the model and the sizes of their domains. If there are N variables, with domains of size $D_1, \ldots D_N$, then the number of solutions is $D_1 \times D_2 \times \ldots \times D_N$.

The instance of the boats and docks problem, with 6 docks and 10 boats is modelled with 10 variables (one for each boat), each have a domain of size 6 (representing the alternative docks to which it could be allocated). The number of candidate solutions for this instance is therefore $6^{10} = 60466176$.

One way to solve this problem is to try each candidate solution in turn, and check whether it satisfies the constraints. This is called the "generate and test" algorithm, and since we will refer to it often, we shall copy a definition from the McGraw-Hill dictionary:

Definition *Generate and test* is a computer problem-solving method in which a sequence of candidate solutions is generated, and each is tested to determine if it is an appropriate solution.   □

This algorithm is very general, and can be used for any problem with a finite number of candidate solutions. This is one reason we suggested that variable domains should be finite in Sect. 2.2.1 above.

However as the number of candidate solutions grows, the time required to perform generate and test grows too. Suppose we could check a million candidate solutions per second, we could go through all the 6-dock 10-boat problem instance in a minute. However for the 12-dock, 20-boat problem, generate and test takes $(12^{20}/1000000)$ which turns out to be more than 100 million years!

5.2.2 Lazy and Eager Generate and Test

We can do better than this if we avoid "lazily" waiting till a candidate solution (i.e. a complete assignment of values to all the variables) has been constructed before we perform the test. We can instead construct candidate solutions by assigning a value to one variable at a time, and "eagerly" testing each new partial assignment immediately.

Let us estimate how long it takes to solve the TSP using generate and test. An N-city TSP instance has N variables each with a domain of size N, so the number of candidate solutions is N^N. The 15-city TSP instance above therefore has 15^{15} which is about 4×10^{17}.

If we test each partial assignment as soon as it is constructed we can immediately rule out any partial assignment in which the same city appears twice. Consequently we have 15 choice for the first variable but only 14 for the second, 13 for the third and so on.

The expression $15 \times 14 \times \ldots \times 2 \times 1$ is called *15 factorial* and written "15!".

The total number of partial assignments generated and tested is therefore no more than 15 factorial which is approximately 10^{12}. For a TSP instance with size N the number of partial solutions generated and tested is $N!$ (N factorial).

We can compare the number of checks done by a lazy generate and test N^N with the number done by eager generate-and-test $N!$, for various values of N: Even though N^N is much bigger than $N!$ it turns out that $N!$ is so big anyway that the difference is not very significant. This is the topic of the next section.

Comparing exponential with factorial growth

N	N^N	$N!$
3	27	6
5	3125	120
7	823,543	5,040
10	10,000,000,000	3,628,800

5.3   The Concept of Algorithmic Complexity

5.3.1   Can We Measure Problem Difficulty?

Algorithmic complexity is the study of how algorithms consume more computational resources as problem instances they are solving grow bigger. Henceforth we will just call it "complexity".

In this text we will focus of just one resource—*computing time* (which is a proxy for the number of steps done by the algorithm.)

We have made the point already that it is hard to estimate how difficult a problem is to solve. The chocolate-breaking problem, and the containers-and-modified-dockside problem, while apparently difficult, turn out to be easy.

Although we can't say how difficult a problem is, we can measure the computing time of algorithms used to solve them. Thus if we can find the "best" algorithm for solving a problem, then the computing time of the algorithm is a good estimate for the difficulty of the problem itself.

There are, unfortunately, two drawbacks to this approach. The first is a show-stopper: given an algorithm for solving a problem, how do we know if it is the best? This question underlies perhaps the deepest outstanding problem in computer science today—"does $P = NP$"?. We will come to this problem in Sect. 5.5.

The second drawback is that of specifying what computing time is needed to run an algorithm. Given the variations in programming languages and compilers, and in computer hardware and architectures, how can we give a "standard" the computing time for an algorithm? Moreover the notion of a standard computing time would be unhelpful because computer performance improves so rapidly from year to year.

5.3.2   Complexity Is a Measure of Scalability

The concept of complexity enables us to get a grip on algorithm performance. The first idea is not to associate a value with the performance of an algorithm on a particular problem instance (see definition in Chap. 2, Sect. 2.2 above) but instead to look at how the algorithm scales with larger and larger problem instances for problems of the same class (see definition in Chap. 2, Sect. 2.2 above). An algorithm scales well if its computing time grows only a little as the size of problem instances is increased. It scales poorly if only a small increase in problem instance size results in a large increase in computing time.

To give a complexity measure for algorithms whose computing time increases with problem size we need to be precise about the size of a problem instance. Assuming one model for a problem class, the size of a problem instance is the result of its parameters.

In some cases the size of a parameter impacts the number of decision variables in the model, and in other cases the size of the parameter itself is what matters. For example in knapsack problem, in Sect. 2.2.4, the number of decision variables is given by the parameter s, which is the number of candidate items for inclusion. The other parameters, k, the size of the knapsack, n the number of items that can be included, and *pitems* the sizes of the items themselves, also impact the size of the problem.

Although the size of a problem is a function of all its parameters values, when defining complexity we typically vary only one of its parameters.

For example if we run the algorithm on a problem instance with parameter N and we run the same algorithm on a problem instance with parameter $2 \times N$ we might expect the algorithm to take twice the amount of time on the second instance.

However when N is the number of decision variables, the generate and test method we discussed above appeared to scale much more poorly than that, both on the boats and docks problem and on the TSP. Doubling the number of boats from 10 to 20 and the number of docks from 6 to 12, increased the solving time from about 1 minute to 100 million years. On the TSP of size N, the number of tests, $N!$, grows by a factor of $N + 1$ when the number of cities increases from N to $N + 1$. Thus the 16-city problem is 16 times harder than the 15 city problem.

It is surprising how much harder the TSP becomes when you add just a few cities. Consider a gradual increase from 15 up to 22 cities. How much harder than the 15-city problem is the 22-city problem? Table 5.2 spells this out.

5.3.3 Complexity Is an Approximation

The second idea is not to give a precise measure of time, or number of steps, for the algorithm, but to give an approximation of its rate of growth with the size of the problem.

In general we say that one algorithm has a *higher* complexity than another if the time used by the first algorithm grows faster than that used by the second.

Consider an algorithm to determine if an input integer is divisible by 7. The parameter is the input integer, and the size of the problem grows as the integer

Table 5.2 Growth in number of possible solutions for increasing TSPs

Number of cities	Calculation	Number of times larger than 15-city problem instance
16	16	16
17	16×17	272
18	$16 \times 17 \times 18$	4,896
19	$16 \times 17 \times 18 \times 19$	93,024
20	$16 \times 17 \times 18 \times 19 \times 20$	1,860,480
21	$16 \times 17 \times 18 \times 19 \times 20 \times 21$	39,070,080
22	$16 \times 17 \times 18 \times 19 \times 20 \times 21 \times 22$	859,541,760

becomes larger. Suppose for example the algorithm performs a division as we are taught at school dividing repeatedly from the left. The number of times we divide increases with the number of digits in the input.

Note, that if 7 divides the number successfully, then the remainder is 0.

1. To determine whether 256 is divisible by 7

   a. Compute $25/7 = 3$ remainder 4
   b. Compute $46/7 = 6$ remainder 4
   c. So 256 is *not* divisible by 7.

2. To determine whether 2568 is divisible by 7

   a. Compute $25/7 = 3$ remainder 4
   b. Compute $46/7 = 6$ remainder 4
   c. Compute $48/7 = 6$ remainder 6
   d. So 2568 is *not* divisible by 7.

3. To determine whether 25683 is divisible by 7

   a. Compute $25/7 = 3$ remainder 4
   b. Compute $46/7 = 6$ remainder 4
   c. Compute $48/7 = 6$ remainder 6
   d. Compute $63/7 = 9$ remainder 0
   e. So 25683 *is* divisible by 7.

In this algorithm the number of iterations—the number of times the process, of dividing by 7, is repeated—increases with the number of digits in the parameter, two for 256, three for 2568, four for 25683 and so on. The complexity of the computation required within each iteration—dividing a small number by 7—is independent of the size of the problem instance.

For the purposes of complexity analysis we do not count the time, or number of steps, performed within each iteration, noting only that it is independent of the size of the problem. For complexity analysis it is, however, significant that the number of iterations increases with the size of the problem.

We now consider an algorithm to determine if an input integer is divisible by 2. This seems a very similar problem to the previous one. How quickly can one determine if the following numbers are divisible by 2?

```
256
2568
25683
256834
```

In this case, we can determine whether a number is divisible by 2 by examining just the last digit, no matter how big the number is. This algorithm (look at the last digit, and answer "yes" if it is even and "no" if it is odd) runs in a time which is independent of the size of the problem.

This algorithm has a lower complexity than our algorithm for dividing by 7, because this algorithm grows more slowly with the size of the problem parameter (in fact it does not grow at all).

5.4 Measuring Complexity

We have used the notions of "higher" and "lower" complexity, and in this section we describe how to measure complexity itself.

The appendix contains additional information about complexity, and how to work out the complexity of an algorithm.

5.4.1 Worst-Case Performance

Clearly an eager generate-and-test, checking each partial assignment, can save a lot of time in a generate and test algorithm. However without knowing in advance which tests are going to fail, we cannot be sure how much time (if any) will be saved. For example when the generate-and-test algorithm happens to construct a complete feasible assignment, then all the checks during the construction of this assignemt will succeed. In this case lazy generate and test is faster than eager generate-and-test.

More generally, there are many algorithms that can terminate quickly if the data happens to be just right. However on a slightly different set of data the algorithm may need a very long time to terminate.

For this reason when reporting the complexity of an algorithm we normally give a "worst-case" complexity, which assumes that everything we don't know about the problem or the algorithm turns out as badly as possible (in daily life this is called "Murphy's law" in America and "Sod's law" in England).

5.4.2 Constant Complexity

The lowest value of complexity is *constant* complexity. An algorithm whose computing time does not increase with the size of the problem has constant complexity. We have already come across an algorithm with constant complexity in the previous section: the algorithm to compute whether a number is even.

5.4.3 Logarithmic Complexity

The algorithm above, for determining whether a number is divisible by 7, requires a number of iterations which increases with the number of digits in the parameter value. The time required by each iteration does not increase with the number of digits. Thus the overall computing time increases at the same rate as the number of digits in the parameter value.

The number of digits increases much more slowly than the value itself. For example as numbers increase from 1, to 10 to 100 to 1000, their number of digits only increases from 1 to 4. We say that the number of digits increases with the "log" of the number itself.

Similarly the time our algorithm requires for computing whether a number is divisible by 7 grows with the log of the number. We call this complexity value "logarithmic" complexity.

Indeed we can draw a graph of the size of a problem against the time required to compute it using an algorithm with logarithmic complexity in Fig. 5.2.

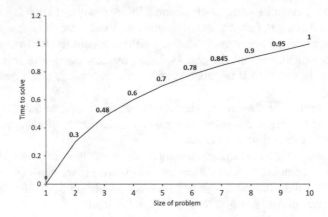

Fig. 5.2 Plot of time against size of logarithmic problem: N

5.4.4 Linear Complexity

Consider the problem of finding the sum of a set of numbers. This problem class has two parameters: the size of the set N (i.e. the number of items it contains) and the set S itself. Let's restrict members of S to be numbers smaller than some given maximum (the maximum can be as large as we want—say 1,000,000).

An algorithm for finding the sum is to start with a sum of 0. Thereafter members of the set are removed one after another, the current member being added each time to the current sum. There is a worst-case time for adding this number to the current sum—for example 6 units of time.

The number of iterations of this algorithm increases by one when the size of the set N increases by one. The amount of time increases by 6 units. We say that this algorithm for finding the sum of the members of a set is *linear* in the parameter N, because when measuring complexity it is the number of iterations that matters and not the precise time. For example one computer may need 6 time units to add the number, but another may need 2 or 10. We use the term "linear" because if we plot the time against the size of N we get a straight line, as in Fig. 5.3. Whatever the number of steps needed to add a number, the slope of the graph is affected, but the complexity remains linear.

We can express the time required to solve a linear problem of size N as $K \times N$ where K is some (possibly unknown) constant.

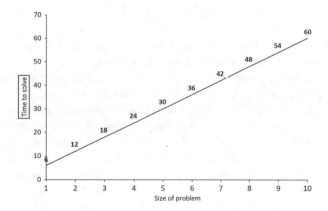

Fig. 5.3 Plot of time against size of linear problem: N

In this problem class, there are two parameters: the size N of the set, and the maximum value M of its members. Because of this, we had to be explicit about what we used to measure the size of the problem instance—in this case N, with M fixed to 1,000,000.

5.4.5   Exponential Complexity

Another algorithm we have already encountered is the generate and test algorithm. When used to solve the boats and docks problem, this algorithm generates each candidate solution in turn and tests it for feasibility. In the worst case, every single candidate solution has to be tested. Suppose we double the number of boats from N to $2 \times N$ but keep the number of docks fixed at some number D. Then the number of candidate solutions increases from D^N to D^{2N}.

Recall D^N is $D \times D \times D \ldots \times D$ where D is multiplied by itself N times. N in this example is the *exponent*.

D^3 is $D \times D \times D$ and

$D^{2 \times 3}$ is $D \times D \times D \times D \times D \times D = (D \times D \times D) \times (D \times D \times D)$.

This illustrates that

$D^{3 \times 2} = (D^3) \times (D^3) = (D^3)^2$,

and in general:

$$D^{2 \times N} = (D^N)^2 \tag{5.1}$$

Thus if we double the number of boats the computing time for this algorithm is *squared*. (If we treble the number of boats, the computing time is *cubed*.)

For an *exponential* problem we can express the time to solve a problem of size N as $K^{p(N)}$, where $K \geq 2$, and $p(N)$ is any expression involving N.[2]

Note that, for any algorithm, if you double the problem size and the computing time is at least squared, its complexity is exponential. Figure 5.4 shows the growth of 2^N as N increases from 1 to 10. We earlier discussed the more efficient eager generate-and-test that tests every partial solution. On the TSP problem the number of tests is (worst-case) $N!$ where N is the number of cities. We can show that $N!$ is exponential just by showing if we double the value of N, the value of $N!$ is more than squared. First we double the size of N to $2 \times N$ and compute $(2 \times N)!$. Note that $(2 \times N)!$ is

$2 \times N \times (2 \times N) - 1 \times (2 \times N) - 2 \times \ldots \times N + 1 \times N \times N - 1 \times \ldots \times 1$

Now we compute $(N!)^2$, which is:

$N \times N - 1 \times N - 2 \times \ldots \times 1 \times N \times N - 1 \ldots \times 1$

---

[2]Depending on the expression the problem may be *sub-exponential*, *exponential* or *double exponential*.

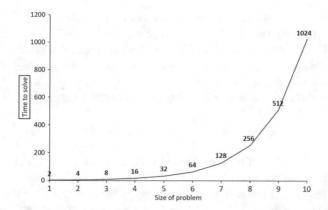

Fig. 5.4 Plot of time against size of exponential problem: N

Since the first N terms in the expression for $(2 \times N)!$ are all larger than the first N terms in the expression for $(N!)^2$, and the rest are the same, we can conclude that $(2 \times N)!$ takes a larger value.

So when we double the number of cities, the computing time is more than squared. From a complexity point of view, then, if both the "lazy" generate and test—where only candidate solutions are tested—and the "eager" generate and test—where every partial assignment is tested immediately—for the TSP have exponential complexity.

Section 9.1 shows that, even though they have the same theoretical complexity, eager generate and test is much better in practice.

5.4.6 Polynomial Complexity

There is a level of complexity between linear and exponential, and indeed many algorithms run on today's computer belong to this class.

Consider the assignment problem of Sect. 2.2.4 in Chap. 2. For n agents and tasks, the number of values in the *value* array is $n \times n$. Thus even the simple task of finding the highest value for assigning one agent to one task requires $n \times n$ values to be checked. In this case the time increases with the *square* of the parameter n. We can also say it increases "quadratically".

Another example is checking whether a set of n values are all different from each other. After checking the first k values, checking the $k+1$th value takes k steps (check the $k+1$th value against each of the previous k values). Therefore checking all n values takes $1+2+\ldots+k+\ldots+n$ steps. This sum is $n \times (n+1)/2$, which also increases with the square of n.

A *polynomial* expression in n is an expression of the form

$$k_m \times n^m + k_{m-1} \times n^{m-1} + \ldots + k_2 \times n^2 + k_1 \times n + k_0 \tag{5.2}$$

where some of the k_i can be zero. Indeed if k_0 is the only one which is non-zero, then the expression is constant: k_0. If k_1 and k_0 are the only ones that are non-zero, then the expression is linear in n. If k_2, k_1 and k_0 are non-zero then the expression is quadratic in n, and if k_3 is also non-zero then the expression is cubic in n.

An algorithm has *polynomial* complexity if its computing time can be expressed as a polynomial in n.

5.5   P and NP

We conclude this chapter by introducing the most famous open problem in computer science. The problem is formulated opaquely as the question "Does $P = NP$?". The complexity class P refers to all problem classes which can be solved by algorithms that run in polynomial time on a conventional computer (also called polynomial, or P, algorithms). The NP stands for "non-deterministic polynomial", and the complexity class NP refers to all problem classes which can be solved by algorithms which run in polynomial time on a particular type of computer which will be introduced shortly. (We also call these non-deterministic polynomial, or NP, algorithms.)

Non-determinism arises when an algorithm is incompletely specified. Indeed this has been the case for several of the algorithms we have discussed in previous sections, including the TSP and summing-the-members-of-a-set. Such algorithms leave open the choice between a finite number of well-specified alternative steps. This is the case for both the example algorithms mentioned above. We call them *nondeterministic* algorithms.

The algorithm for finding the sum of the members of a set, for example, removes members of the set one-at-a-time, but the algorithm does not specify *which* member is removed each time.

Neither the lazy nor eager generate and test algorithms for the TSP specify which in which order the candidates should be tested. For the eager generate and test algorithm we can diagram the search space as a tree shown in Fig 5.5.

The algorithm explores the nodes of this tree starting from the "root" (misleadingly at the *top* of the tree) and progressing down each branch of the tree, testing each partial assignment as it goes. The order in which the branches are explored

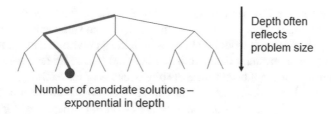

Number of candidate solutions –
exponential in depth

Fig. 5.5 TSP—tree of partial assignments

are not specified. If the test succeeds on a "leaf" of the tree (corresponding to an assignment for all the variables, and thus a candidate solution) then this is a solution.

A non-deterministic machine is one that runs a nondeterministc algorithm in the best way possible. It can be viewed as always making the "right" choice that leads most efficiently to a solution. In the above figure if the green branch leads to the required solution, the nondeterministic machine simply follows it. A nondeterministic machine can also be viewed as making all the alternative choices at the same time so the right choice is just one of them.[3]

In any case the time taken to run the algorithm on a nondetermistic machine would appear to be shorter than the worst-case complexity of running it on a normal machine, where by Murphy's law the machine always makes the *wrong* choice! Indeed on a TSP with N cities, the nondeterministic machine need only check N partial solutions on the way down the correct branch. However as we saw earlier, the worst-case eager generate and test algorithm makes $N!$ tests. (Notice that for finding the sum of the members of a set, the nondeterministic algorithm still has to handle every member of the set—there are no choices to make, and so it requires just as many steps as the deterministic algorithm.)

An *NP* algorithm is one that runs in polynomial time on a nondetermistic machine. This is the case for the TSP, assuming the test on each partial assignment itself takes polynomial time. As we saw above a linear number of iterations on a polynomial subproblem yields a polynomial iterative algorithm.

The open question "Does $P = NP$?" asks if every problem class which can be solved by an *NP* algorithm can also be solved by a polynomial algorithm?

At its heart, the question if $P = NP$ is a question about practical computability. If the answer is yes, which means that industrial planning, scheduling and resource allocation problems can be solved in polynomial time, then it would be possible for an out-of-the-box algorithm to find optimal solutions to many of these problems.

It has been a goal of quantum computing to be able to solve problems of this class in polynomial time, though the theory has not so far established that this goal is achievable. In some sense, however, if $P = NP$ much of the wind would be taken out of the sails of quantum computing.

---

[3] Although quantum computers can indeed represent multiple states simultaneously, they cannot unfortunately behave as nondeterministic machines.

Examples like the TSP seem to suggest that $P = NP$ is false: the decision variant of the TSP (asking if there is a circuit with cost less than some value) can be solved by an *NP* algorithm, but it would be hard to imagine a polynomial algorithm (worst-case, running on an ordinary deterministic machine) that could also solve it. Indeed such an algorithm has not been found despite decades of research.

Nevertheless it has yet to be proven that such an algorithm does not exist and, moreover, the validity of the equation $P = NP$ is yet to be disproved (or proven).

5.6 Problem Complexity

5.6.1 Algorithms for Solving IDSS Applications

Exponential Algorithms

In Sect. 5.4.5 we have seen how the number of candidate solutions increases exponentially with the number of decision variables. We have also defined the complexity of an algorithm, and shown that the generate and test algorithm, which tests all candidate solutions to a problem, has exponential complexity. By this we mean that, even if all the variable domains have the same size (say D), if the number of variables is V, then the number of candidate solutions D^V increases exponentially with V.

The real challenge of IDSS is that the algorithms we would have to use to find optimal solutions, and guarantee their optimality, are very often exponential. We are therefore required to apply our judgement in balancing the quality of the solution found against the computational resources—and in particular the computer running time—required to find it. Significantly it is extremely difficult to guarantee that the "good" solution found by the chosen algorithm is anywhere near optimal. Indeed a much better solution might have been quickly discovered by another algorithm.

Optimally Locating a Warehouse

Nevertheless there are problems which can be solved by algorithms with a lower complexity. In cases where such an algorithm is available it is important to employ it! Consider, for example, a warehouse location problem

- Place one warehouse
- in one of a finite set of locations
- minimising cost of transport from warehouse to customers
- with no limit on the size of the warehouse

To find the best solution, we try locating the warehouse in each location in turn, and compute the total cost of transport from all the customers to that locations. The warehouse location with the lowest total cost is the optimum solution. In this case if there are L possible warehouse locations and C customers, we merely add together C transport costs for each of the L locations. The algorithmic complexity is simply $L \times C$ which is polynomial in L (if C remains the same), and polynomial in C (if

L remains the same). For any realistic values of L and C, all L candidate solutions can be quickly checked and the optimum solution found.

Complexity of the TSP

The travelling salesman problem (TSP) requires C locations to be visited by the shortest possible tour. The TSP has been tackled by many researchers but the best algorithms still have complexity exponential in C. There are many varieties of the TSP (TSP with time windows, TSP with pickup and delivery with limited capacity vehicles etc.) which are all hard to solve. An IDSS salesperson who claims to have an algorithm which solves to optimality medium to large scale industrial applications involving pickups and deliveries (e.g. mail, parcels, home deliveries, security vans delivering money, etc.) is probably unaware that researchers have for years failed to find an algorithm that could do this, and does not understand the limitations of the software he or she is selling.

Production Scheduling

Scheduling production tasks on one or more machines is an interesting case. In the famous "job-shop" problem each job comprises a sequence of tasks which must be carried out one after another on different machines with the last task ending as early as possible, or within a given (tight) deadline. It is hard to solve this problem. All algorithms found to date are exponential in the number of tasks, and for problems with over 250 tasks finding a guaranteed optimal solution is beyond the state of the art. Highly parallel computing can increase that number by maybe 10 or 20 tasks, but by the nature of exponential algorithms parallelism cannot help increase the number of tasks by much.

If there is only one machine, and all the tasks have a possibly different due date, then we can solve the problem efficiently, by simply sorting the tasks by earliest due date and processing them in due-date sequence. Using this algorithm problems with millions of tasks can be solved! On the other hand if each task has both a due date and a *release* date (an earliest starting time), the best algorithms are again exponential in the number of tasks.

As these scheduling problems show, it is challenging to develop an IDSS because it is hard to know whether the algorithm selected for the problem is indeed the best one for the problem class.

5.6.2 NP Complete Problems

In Sect. 5.5 we investigated the complexity class *NP* which can be solved by algorithms that run in polynomial time on non-deterministic computers. Many problem classes, such as the decision variant of the TSP, belong to the complexity class *NP*. Moreover, as mentioned above, no polynomial complexity algorithm (running on a normal computer) has been found to solve the TSP.

The open question presented in Sect. 5.5 "Does $P = NP$?" is a question about problem classes, rather than algorithms. It asks if there are problem classes which can be solved by NP algorithms but not by any polynomial algorithm.

To follow this up in more detail, any problem in NP can be solved in polynomial time on a "non-deterministic" computer. For any given input and state, a non-determistic computer changes to a new state: this is no different from a normal, deterministic, computer. Hower in a non-deterministic computer, this new state may be any one of a finite set of alternative states. Note that for any problem in NP the compilation of the problem into code that runs on the non-deterministic machine itself runs in polynomial time.

Stephen Cook proved in 1971 that there is a problem class which represents all problems in the class NP. The class is called "SAT", and we will learn more about it later. Cook showed that the behaviour of any non-deterministic machine can be encoded as a SAT problem. If, for a given input, the machine reaches a terminal (success) state in a polynomial number of steps, then the size of the SAT problem model is polynomial. The (polynomial) sequence of states of the non-deterministic machine, ending in the terminal state, translates to a solution of the SAT instance. Thus any terminating polynomial algorithm on this machine can be encoded as a solution of the SAT instance. In short every NP algorithm corresponds to a solution of a SAT problem.

By finding an algorithm for problems of this class on the non-deterministic machine, and the formalising the machine's behaviour as a SAT model, we can map every NP problem instance to a SAT problem instance in a time which is polynomial in the size of the instance. The process of mapping an original problem to the resulting SAT problem is termed *reduction*. We say the original problem has a polynomial reduction to SAT. Since any problem in NP has a polynomial reduction to SAT we say that SAT is *NP-complete*. It turns out that many important problem classes in NP are *NP*-complete, including for example the decision variant of TSP.

Definition In general a problem class is *NP-complete* if it is in the class NP and any problem class in NP can be reduced to it in polynomial time.   □

If somebody were to discover a *polynomial* algorithm for SAT, then any NP problem class could be translated to SAT and solved in polynomial time on a conventional machine. Consequently the complexity class NP would be shown to be the same as the polynomial complexity class P. But despite intensive research, no such polynomial algorithm has been found. In this case the IDSS developer can be confident that if his algorithm cannot guarantee to find an optimal solution in a reasonable time,[4] he is in good company!

---

[4]I.e. polynomial in the size of the problem instance.

5.6.3 *Complexity Class* **NP-hard**

The last complexity class discussed in this section is, in principle, a bit harder than the *NP*-complete class. A problem in the class *NP*-hard is not (necessarily) in the class *NP*. For example an optimisation problem is not in the class *NP*, but may be *NP*-hard.

Definition In general a problem class is *NP*-hard if any problem class in *NP* can be reduced to it in polynomial time.   □

Equivalently, a problem is *NP*-hard if an *NP-complete* problem is reducible to it in polynomial time. Informally this means that to solve an *NP*-hard problem it is necessary to solve an *NP*-complete problem on the way.

A typical example of an *NP*-hard problem is the TSP *optimisation* variant. This variant of the problem requires an *optimal* tour to be found, rather than simply a tour whose length is within a given bound. Clearly this problem is at least as hard as the decision variant of the TSP, because finding an optimal tour—and in particular its length—immediately answers all the decision problems of the form "Is there a tour with length less than T?". (If the optimal length is less than T the answer is "yes" and otherwise the answer is "no".)

The final step in the TSP optimisation problem—guaranteeing that the best solution found so far is indeed an optimal solution—is to solve the decision variant of the TSP problem. This step, if the best solution found so far has length Opt, is to answer the question "Is there a tour whose length is less than Opt?" Many decision problems in the class *NP* have optimisation variants, and these all belong to the *NP*-hard class.

5.6.4 *Relating the Different Complexity Classes*

The polynomial complexity class P is a subclass of *NP* because every problem class whose complexity is P can also be solved by the same polynomial algorithm on a non-determistic machine.

The class *NP* may also be the same as the class P if somebody one day finds a polynomial algorithm for solving every problem in an *NP*-complete class, such as the SAT class.

The class of *NP*-complete problems is also a subclass of *NP* (by definition any *NP*-complete problem is in the class *NP*), but if $P \neq NP$ then there are also problems in *NP* that are not *NP*-complete.

Finally the *NP*-hard problem class is a superclass of *NP*. Every *NP*-complete problem is *NP*-hard (because by definition every problem in *NP* is polynomially reduceable to it), but many *NP* hard problems are not in the class *NP*, so they are not *NP*-complete either.

We conclude with Fig. 5.6 relating all these complexity classes.

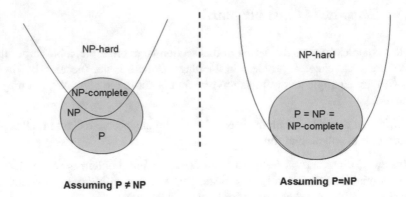

Fig. 5.6 Complexity classes

5.7   Summary

For a given problem instance, by ignoring all the constraints, it is possible to count the candidate solutions. The number depends on the number of decision variables and the sizes of their domains. The classic generate-and-test algorithm checks all the candidate solutions, so the computational effort grows very fast with the size of the problem instance.

It is hard to measure the computational cost of an algorithm in a way that is independent of the computer it runs on. Computational complexity achieves this by measuring how many more iterations an algorithm needs for larger problem instances of the same class. An *iteration* is any sequence of steps that is independent of the size of the instance. The computational complexity of an algorithm may be *constant* in the best case, linear, polynomial, exponential, or worse. This measure assumes the data is not chosen to make the instance easy to solve: we term it *worst case* complexity. The complexity of a problem class is determined by the complexity of the most efficient algorithm for solving it.

A variant of complexity is *non-deterministic* complexity, which represents the cost of an algorithm running on a theoretical non-deterministic machine. Many problems whose best known algorithm is exponential have non-deterministic polynomial (or *NP*) complexity. It is an open question whether any problem solved by an *NP* algorithm can also be solved by a polynomial algorithm: does $P = NP$? However certain problem classes, termed *NP-complete* have been shown to represent all *NP* problem classes.

Chapter 6
Constraint Classes

In this chapter we will investigate the different kinds of constraints that can occur in problem models. In particular we will investigate models that contain only certain kinds of constraints. The kinds of constraints that appear in a model dictate how hard it may be to solve. To help understand this relationship we introduce a number of *constraint classes*, and discuss what kinds of problems can be modelled using the different constraint classes.

6.1 Introduction

6.1.1 Five Classes of Constraints

A constraint (in MiniZinc and more generally) is a logical expression that must be *true*. $X > 2$ is an expression whose value may be *true* or *false* depending on the value of X. If, however, $X > 2$ is imposed as a constraint, then X is only allowed to take values which make $X > 2$ true.

We note firstly that a constraint has a number of variables in its scope – in the above example, the single variable X. The types of these variables are dictated by the constraint. We could not (meaningfully) write "John" > 2 since *"John"* is not a number. (We sometimes say the *type* of X, in $X > 2$, is *number*.)

Similarly the type of expressions appearing in the constraint are dictated by the constraint class. Thus only certain functions and connectives may be allowed. For example $X < 2 + 3$ is a constraint in which the expression $2 + 3$ appears, and the function $+$ has to be a function that returns a number.

By restricting the class of constraints that can appear in a model, we also restrict the types of its variables, and expressions.

© Springer Nature Switzerland AG 2020
M. Wallace, *Building Decision Support Systems*,
https://doi.org/10.1007/978-3-030-41732-1_6

The constraint classes that will be explored in this chapter are:

- Finite domain constraints
- Propositional (logic) constraints
- Linear constraints, floating point variables and linear expressions
- Integer linear constraints, with associated numerical variables and expressions
- Non-linear constraints, with unrestricted numerical variables and expressions

6.2 Finite Domain Constraints

6.2.1 Finitely Many Combinations to Check

Finite domain constraints are constraints over expressions which themselves can only take finitely many alternative values. Many of the problems we encounter have only a finite set of possible choices, including all the examples from Chap. 2.

To solve such problems it suffices to check every combination of values for the decision variables, and for each combination of values, check all the constraints are feasible. There can be a lot of combinations, of course, but the process is guaranteed to correctly check every combination of variables. This is illustrated in two examples from Chap. 2.

Knapsack Problem

For example in the knapsack problem of Sect. 2.2.4 above, there are 8 variables in the array $PVars$. Given a combination of values for the decision variables, such as $0, 0, 0, 0, 1, 1, 1, 1$, there are two constraints to be checked. The first is:

$$\sum PVars = 5$$

which for these values becomes:

$$0 + 0 + 0 + 0 + 1 + 1 + 1 + 1 = 5$$

This is false. There is no need to check the second constraint for these values, since they do not satisfy the first constraint, but for completeness we show the checking of the second constraint:

$$\sum_{j=1..S} (pitems[j] \times PVars[j]) = 58$$

which for these values becomes:

$$5 \times 0 + 7 \times 0 + 9 \times 0 + 10 \times 0 + 14 \times 1 + 16 \times 1 + 17 \times 1 + 18 \times 1 = 58$$

(also false).

Table 6.1 Assignment values

value =	1	2	3
1	2	2	3
2	4	4	4
3	7	6	5

Assignment Problem

The assignment problem on page 23 above has an array A of 3 variables, and an array TV of 3 variables. Suppose limit = 12 and the matrix of assignment values, *value*, is input as Table 6.1. A possible combination of values for the decision variables is

$$A[1] = 1, A[2] = 2, A[3] = 3, TV[1] = 4, TV[2] = 4, TV[3] = 4,$$

There are three constraints:

$$A[x] \neq A[y] : x \in 1..2, y \in x + 1..3$$

which for these values becomes true:

$$1 \neq 2, 1 \neq 3, 2 \neq 3$$

The constraint

$$t \in 1 \ldots 3, TV[t] = value[A[t], t]$$

which becomes false:

$$4 = 2, 4 = 4, 4 = 5$$

and

$$\sum TV \geq limit$$

which becomes true:

$$4 + 4 + 4 \geq 12$$

However because the second constraint is unsatisfied, this is not a feasible solution.

6.2.2   *Logical Combinations of Constraints*

Suppose in the Boats and Docks problem above (Sect. 2.2.3), the harbourmaster said "Either boat b1 must go in dock4 or b3 must go in dock1". (This might be related to some prior unloading agreement.) We can't write two constraints:

$$Boat1 = dock4$$

$$Boat3 = dock1$$

because this would impose that both b1 goes in dock4 *and* b3 goes in dock1. So we use the logical connective *or* and write the constraint as:

$$Boat1 = dock4 \text{ or } Boat3 = dock1$$

We term constraints with an "or", *disjunctions*.

Similarly we can combine constraints with an "and". Instead of using two constraints to say "b3 must go in dock1 and b4 must go in dock2", we can use one constraint and combine the two expressions with the logical connective *and*, writing:

$$Boat3 = dock1 \text{ and } Boat4 = dock2$$

A constraint with "and" is termed a *conjunction*

Another possible requirements might be: "It is not possible for both b1 and b2 to go in dock4". This can be expressed using *not*, and we write:

$$\text{not } (Boat1 = dock4 \text{ and } Boat2 = dock4)$$

This is a *negation*. The brackets are important here, because the constraint:

$$(\text{not } Boat1 = dock4) \text{ and } Boat2 = dock4$$

means something different.[1]

Finally a third requirement might be "If b1 goes in dock4 then b2 goes in dock3". This can be expressed using *implies*, and we write:

$$Boat1 = dock4 \text{ implies } Boat2 = dock3$$

This is termed an *implication*.

We noted before that a constraint is simply an expression that can take the values *true* or *false* depending on the values of its variables. We term such an expression

---

[1]It means "b1 cannot go in dock4, and b2 *must* go in dock4".

Table 6.2 Truth table for *and*

B1	B2	B1 and B2
True	True	True
True	False	False
False	True	False
False	False	False

Table 6.3 Truth table for the propositional connectives

B1	B2	B1 and B2	B1 or B2	B1 implies B2	not B1
True	True	True	True	True	False
True	False	False	True	False	False
False	True	False	True	True	True
False	False	False	False	True	True

a *boolean* expression, after George Boole whose book "An Investigation of the Laws of Thought [9] laid the foundations of the information age! There can also be boolean variables: a boolean variable is one with domain {*true, false*}.

If B is a boolean variable, we can write $(X > 2) = B$, which means that if X is greater than 2 then $B = true$, otherwise $B = false$.

$X > 2$ means the same as $(X > 2) = true$.

The logical connectives are best understood in terms of their effect on booleans. For example *not* $(X > 2)$ means the same as $(X > 2) = false$.

The logic of the connective *and* is (obviously) that

$$Boat3 = dock1 \text{ and } Boat4 = dock2$$

means $(Boat3 = dock1) = true$ and $(Boat4 = dock2) = true$. Clearly if either $(Boat3 = dock1) = false$ or $(Boat4 = dock2) = false$ the conjunction is false.

This (obvious) meaning can be written as a *truth table*, as in Table 6.2. $B1$ and $B2$ in the truth table can be thought of as boolean variables or as expressions like $X > 2$ and $Boat3 = dock1$.

In the same way we can write truth tables for *not*, *or* and *implies*, in Table 6.3. Although the truth tables for *and* and *not* are obvious, those for *or* and for *implies* are less so.[2]

Note that by this definition "It is not possible for both b1 and b2 to go in dock4" has the same interpretation as "Either b1 does not go in dock4 or b2 does not go in dock4". In other words

$$not (Boat1 = dock4 \text{ and } Boat2 = dock4)$$

---

[2]Indeed philosophers debate about the meaning of implication when the antecedent is *false* [8].

Table 6.4 X is (Boat1=dock4); Y is (Boat2=dock4)

X	Y	X and Y	not (X and Y)	not X	not Y	(not X) or (not Y)
True	True	True	False	False	False	False
True	False	False	True	False	True	True
False	True	False	True	True	False	True
False	False	False	True	True	True	True

is *true* under exactly the same conditions that make

$$(\text{not } Boat1 = dock4) \text{ or } (\text{not} Boat2 = dock4)$$

true as well.

Any propositional formula brings its own truth table, built from the truth tables of its logical connectives. For example the truth table for
not (*Boat1* = *dock4* and *Boat2* = *dock4*)
and for
(not *Boat1* = *dock4*) or (not*Boat2* = *dock4*)
are in Table 6.4.

In order to evaluate a logical combination of finite domain constraints, it is still possible to try every combination of values, except that now we need to evaluate every combination for every row of the associated truth table.

6.3   Propositional Constraints

This leads us into a new class of constraints. While propositional constraints are also finite domain constraints, the restriction to just the boolean domain *true* and *false*, makes propositional constraints interesting in themselves.

Propositional constraints only admit expressions with boolean variables. In fact any boolean variable is a propositional constraint, and so are the boolean constants *true* and *false*—though the constraint *false* can never be satisfied, so it is not a very useful constraint!

Propositional constraints can be built from boolean variables and constants using the logical connectives *and*, *not*, *or* and *implies*. They can be surprisingly complicated.

Consider the following situation. There are two politicians "Julie" and "Anthony", and both of them make a statement. The problem is to find out who is telling the truth (i.e. whose statement is true) and who is not.

The statements are:

Julie   Exactly one of us is telling the truth
Anthony   At least one of us is telling the truth

Note that if Julie is telling the truth then only one of Julie or Anthony is telling the truth, but not both. On the other hand, if Anthony is telling the truth, then either Julie is telling the truth, or Anthony is, or both are.[3]

We model this problem using two boolean variables J (for Julie) and A (for Anthony). J takes the value *true* if and only if Julie is telling the truth (i.e. $J = false$ if Julie is lying), and A takes the value *true* if and only if Anthony is telling the truth.

The statement "Exactly one of us is telling the truth" is modelled by the boolean expression "$(J$ or $A)$ and not $(J$ and $A)$". According to the problem specification this expression is *true* if and only if Julie is telling the truth, i.e. J is *true*. We therefore have the first constraint

$$J = ((J \text{ or } A) \text{ and not } (J \text{ and } A))$$

The second constraint is similar:

$$A = (J \text{ or } A)$$

It is instructive to try and work out whether these two constraints are both satisfiable at the same time, and if so could Julie be telling the truth?

Propositional models can be very large and complicated. Indeed, at the level of integrated circuits, every computer program is actually represented as a propositional model. However, in an integrated circuit, the input to each gate (i.e. each propositional connective) is given, and the output is immediately (and very efficiently) computed from this input. Each gate works in only one direction—from input to output. Propositional models in general are much more difficult to solve, because consistent values of all the variables (input and output) must be found to satisfy all the constraints.

6.4   Linear Constraints

6.4.1   Floating Point Expressions

Many practical problems require us to reason over other numbers than just integers.

Confectionary Model
We start with an example of manufacturing confectionary from a set of raw materials including butter and sugar. Suppose we are manufacturing two kinds of confectionary, butterscotch and toffee. The ratio of butter to sugar is different in butterscotch and toffee. Also the value of the finished confectionary, per unit weight,

---

[3]"Either..or.." in English is typically meant as the *exclusive or*, but in logic "or" means the *inclusive or*.

Table 6.5 Confectionary: composition and value

	Butter	Sugar	Value
Butterscotch	2	1	5
Toffee	2	3	8
Qty available	10	10	

is different for the two kinds of confectionary. Let us summarise this information in a table (Table 6.5). The question is, given the limited amount of butter and sugar available, what is the maximum value of confectionary that can be produced?

The maximum value is in fact 32.5, achieved by making 2.5 units of butterscotch and 2.5 units of toffee. This combination uses all the available sugar, and all the butter.

By contrast if we specify the problem using a finite integer model the best solution found yields a lower profit. Specifically the best integer solution produces 3 units of butterscotch and only 2 units of toffee, yielding a total value of 31.

To model this problem in such a way that the optimal, non-integer, solution can be returned, a new class of numeric variables and expressions is required, which can take non-integer values. The non-integer numbers that can be represented on a computer we call "floating point" numbers, or just "floats".[4]

Floating point numbers are usually written with a decimal point, so the model of the above problem using float variables and expressions is thus:

Parameters   – Quantity of Butter: $butter = 10.0$
             – Quantity of Sugar: $sugar = 10.0$
             – Matrix recording the amounts of butter and sugar per unit of butterscotch and toffee:

$m =$	1	2
1	2.0	1.0
2	2.0	3.0

             – Value of butterscotch, per unit: $valB = 5.0$
             – Value of toffee, per unit: $valT = 8.0$
Variables   – Quantity of butterscotch produced: QB
             Domain QB is 0.0 .. 10.0
             – Quantity of toffee produced: QT
             Domain QT is 0.0 .. 10.0
             – Total value of sweets produced: $Value$

---

[4]Note that floating point numbers cannot exactly represent all *real* numbers, and so the answer returned from the solver may not be precisely the *real* number expected by the user. For example the floating point value of X satisfying $3 \times X = 1.0$ returned by the linear solver is $X = 0.3333333333333333$, which is not (quite) the *real* solution.

Constraints   – $m[1, 1] \times QB + m[2, 1] \times QT \leq butter$
               – $m[1, 2] \times QB + m[2, 2] \times QT \leq sugar$
               – $Value = valB \times QB + valT \times QT$
Objective   – Maximise $Value$

6.4.2   Expressive Power and Complexity of Floating Point Constraints

With finite integer variables we have seen that a strict disequality $X > Y$ can be expressed as a non-strict inequality $X \geq Y + 1$. By contrast this is not possible with floating point variables and expressions. In this case $X > Y$ is clearly not the same as $X \geq Y + 1$. For example $4 > 3.5$ is true but $4 \geq 3.5 + 1$ is false. In fact none of the comparisons $>$, $<$ or $\neq$ are expressible as constraints on floats. As a consequence, the negation of a floating point constraint is not expressible as a floating point expression. For example not $X \geq 2.0$ means the same as $X < 2.0$ which cannot be handled.

The consequence is that logical combinations of floating point constraints are not in the class of linear constraints.

6.4.3   Linear Expressions and Constraints

Models involving floating point numbers are fundamentally different from finite integer models because the number of alternative values that a variable can take is so large it is for practical purposes unbounded. Consequently there is no generate-and-test algorithm that can try all possible combinations of values and test them for feasibility.

Surprisingly, linear constraints are in general easier to solve than finite integer constraints.

Efficient techniques for solving models using these constraints have been studied and used in industry for over 60 years. This area of research and application is called "linear programming", and models involving only linear floating point expressions are termed *linear programming* (LP) models.

We now introduce the class of linear constraints through three definitions: *linear term*, *linear expression* and *linear constraint*.

Definition A *linear term* is either a constant c, or a variable V, or the product of a constant and a variable $c \times V$.   □

In a linear floating point term, the constant is a floating point number and the variable a float variable. A simple example is the term $valT \times QT$ in the above model for sweets, since *valT* is a floating point number, and *QT* is a floating point variable.

Definition A *linear expression* is a sum of linear terms.   □

A simple example is the expression *valB* × *QB* + *valT* × *QT* in the same model.

Definition A *linear constraint* is an equation *Expr1* = *Expr2*, or inequality *Expr1* ≥ *Expr2* between two linear expressions *Expr1* and *Expr2*.   □

Note that *Expr1* ≤ *Expr2* is the same as 0 − *Expr1* ≥ 0 − *Expr2*, which is also (therefore) a linear constraint.

The class of linear constraints arise in a wide variety of problems, including production planning as exemplified above.

Transportation Model

Another nice example of modelling with linear constraints is for transportation problems. A simple transportation problem has a set of locations from which goods have to be picked up, and a set of locations where they have to be delivered (see Fig. 6.1). We can portray the pick-up locations as "plants" and the delivery locations as "clients".

Each client has a *demand* which has to be satisfied, and each plant has a *capacity* which it cannot exceed. Finally there is a (different) cost in going from each plant to each client. To model this problem we introduce a decision variable for each link from a plant to a client, whose value is the flow down that link. The constraints say that the total flow out from each plant must be less than its capacity, and the total flow into each client must satisfy its demand. (We allow this flow to be greater than or equal to the demand.)

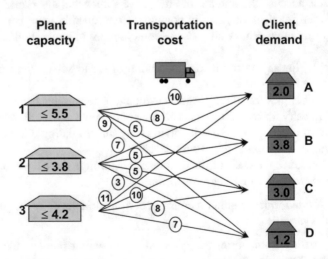

Fig. 6.1 A transportation problem

We can name the variable on the link from plant 1 to client A as X_{1A}, from plant 1 to client B as X_{1B}, etc. Our model therefore has 12 variables $X_{1A} \ldots X_{3D}$ and 7 constraints. The cost of a link is its (input) transportation cost, times its flow. The total cost on all links has to be minimised. The complete (linear) model is as follows:

Parameters   – Array of demands $dem = [2.0, 3.8, 3.0, 1.2]$
             – Array of capacities $cap = [5.5, 3.8, 4.2]$
             – 4 by 3 matrix recording the cost of delivering to each client from each plant:

$cost =$	1	2	3
1	10.0	7.0	11.0
2	8.0	5.0	11.0
3	5.0	5.0	8.0
4	9.0	3.0	7.0

Variables   – 4 by 3 matrix F of variables denoting how much of the demand from each client is supplied from each plant
Constraints – $F[i, j] \geq 0 : i \in 1..4, j \in 1..3$
             – $\sum_{j \in 1..3} F[i, j] \geq dem[i] : i \in 1..4$
             – $\sum_{i \in 1..3} F[i, j] \leq cap[j] : j \in 1..3$
Objective   Maximise $\sum_{i \in 1..4, j \in 1..3} F[i, j] \times cost[i, j]$

For a linear model the objective expression must be linear, like all the other expressions.

6.5   Linear Constraints over Integers and Floats

The "linear programming" (LP) class comprises problems that are modelled using only linear expressions over floating point variables. Problems can also be modelled using linear expressions over both integer variables and floating point variables. These models are termed *Mixed Integer Programming* (MIP) models. The class of MIP models therefore looks very similar to the class of LP models, but MIP models are in general much harder to solve.

A typical variant of the transport problem introduced above, is the problem with the added constraint that a client can only be served from a single plant. This makes sense since the number of available trucks is usually limited.

To model this constraint we would like to impose that of all the flows from the different plants to one client only one is non-zero. Until it is known which of the flows is non-zero, however, this cannot be expressed as a linear constraint on continuous variables.

Using a zero-one variable for each plant-client link, as well as the original flow variable, it becomes possible to model this constraint. To start, we need an upper bound on the flow between the plant and the client: two obvious upper bounds are the capacity of the plant or the demand of the client. We now have a floating point number UBF which is the upper bound on the flow, a continuous variable F for the flow, and a zero-one variable B to allow or disallow a non-zero flow on this link. The constraint $F \leq B \times UBF$ has precisely the required behaviour. If $B = 0$ the flow F can only be zero. If, on the other hand $B = 1$ the flow can be anything up to its upper bound, which cannot be exceeded anyway, so this constraint cannot be violated. If $B_1 \ldots B_k$ are the zero-one variables for all the links from different plants into one client, the constraint $\sum_{j \in 1..k} B_j = 1$ captures exactly the constraint that only one of the plants can supply this client.

We extend the model for the original transportation problem in the previous section to a model for the one-plant-per-client variant as follows.

Variables   4 by 3 matrix of zero-one variables, one for each customer and plant MB

Constraints – $F[i, j] \leq MB[i, j] \times dem[i] : i \in 1..4, j \in 1..3$
         – $\sum_{j \in 1..3} MB[i, j] = 1 : i \in 1..4$

Running this problem in MiniZinc (see Sect. 8.5.5), we will see that the new optimal solution has an optimal cost which is somewhat higher than the original transport problem.

Of course industrial production and transport planning problems have many more constraints and variables than these examples, but the same modelling principles apply, in theory and in practice!

6.6 Nonlinear Constraints

The nonlinear constraints discussed in this section are the ones that are not efficiently modelled in other constraint classes. These involve nonlinear expressions over real numbers, modelled as float constants and variables.

6.6.1 Nonlinear Functions and Expressions

A common form of nonlinear expression is a polynomial, in one or more variables, such as $X^2 \times Y + 3.2 \times X \times Y + 7.5$, where X and Y are float variables. A related form of nonlinearity is division by a variable (e.g. $1/X$), which has an added complication when the variable is close to zero. The expression $1/X$ can jump from large and positive when X is positive and close to zero, to large and negative when X is negative and close to zero. Trigonometry introduces another class of nonlinear functions such as sin, cos and tan. Finally when a variable Y appears as an exponent (e.g. 3^Y) the resulting expression is highly nonlinear. There is no formal definition, but informally, a *highly nonlinear* expression is one for which any linear

approximation, even if it is exact at a certain point, quickly diverges from the value
of the expression with increasing distance from this point.

6.6.2 Nonlinear Problems

Nonlinear constraints arise in chemical processing, robotics, financial modelling,
control optimisation, and the modelling of physical systems involving: time, speed,
force, and volume for example; and the modelling of electrical systems involving:
power, resistance, capacitance, inductance, etc.

Since the algorithms currently available for solving non-linear constraints are not
very scalable, a common approach is to approximate them with linear constraints.
This approximation is accurate at a single point and (typically) diverges more
and more with increasing distance from the point. The linear approximation at
a point (a candidate solution) suggests a direction towards better solutions. This
enables the algorithm to choose a (hopefully) better point and make another linear
approximation at this new point. In this way a non-linear problem can be tackled
by running a sequence of linear problems until the final solution appears to be good
enough.

In this section, however, we model the nonlinear problem itself.

Compost Bin

A simple nonlinear problem arises in buying a sheet of wire mesh to wind into a
circle to make a (compost) bin. The width of the sheet is the height of the bin, and
the length of the sheet is the circumference of the bin (see Fig. 6.2). If I need a bin
with size 2 cubic metres, and the mesh comes in a width of 2 metres, what length
of mesh do I need? A more difficult (if less practical) problem arises if the width
of the mesh is also a variable, and the requirement is to minimize the area of mesh
required. In this case we add a constraint that the width and length of the mesh must
be more than half a metre, to avoid impractical solutions.

This is a simple nonlinear problem since the volume V of the bin is a function
of its radius R and its height. The height is simply the width W of the mesh. The

Fig. 6.2 A bin from a length
of wire mesh

circumference of the bin $2\pi R$ is the length L of the mesh. The problem can be simply modelled in thus:

Parameters	–	The value of π: pi = 3.14159
	–	Bin size: $binSize = 2.0$
Variables	–	Mesh width: $MeshW$
		Domain $MeshW$ is 0.5 .. 3.0
	–	Mesh Length : $MeshL$
		Domain $MeshL$ is 0.5 .. 10.0
	–	Bin radius: $BinR$
		Domain $BinR$ is 0.0 ... 10.0
Constraints	–	$binSize = pi \times BinR^2 \times MeshW$
	–	$MeshL = 2 \times pi \times BinR$

At the risk of stating the obvious, it is true that most programming languages allow complex numerical functions to be evaluated. Given a width and length for the mesh, therefore, most programming languages allow the volume of the bin to be computed. One of the challenges in handling constraints is also to discover inputs to numerical functions that yield a required output.

Black-Scholes

Our next example of a nonlinear constraint is the Black-Scholes option-trading model introduced in Sect. 1.3.

$$OP = SP \times n(D1) - StrP \times e^{RFR \times MT} \times n(D2)$$

$$D1 = \frac{ln(SP/StrP) + (RFR + 0.5 \times SV^2) \times T}{SV \times \sqrt{MT}} \tag{6.1}$$

$$D2 = D1 - SV \times \sqrt{MT}$$

where:

- Call option price : OP
- Current stock price: SP
- Exercise price: $StrP$
- Short-term (risk-free) interest rate: RFR
- Time remaining to expiration date: MT
- Standard deviation of the stock price: SV
- Option expiry variable: $D1$
- Option expiry variable: $D2$
- n(.) is the cumulative normal probability

It is a complex equation, but given values for all the variables—SP, $StrP$, RFR, MT, SV—there are many computer systems which can compute the output call option price.

Encoding the equation as a constraint dramatically extends this functionality, by allowing the user to aim for a specific call option price OP by using a nonlinear constraint solver to find a compatible combination of exercise price *StrP* and time to expiration date *MT*, for example. Of course this constraint can occur in a larger financial model which might impose further constraints on *StrP* and *MT* in combination with other decision variables. Here, as a more interesting example of nonlinear constraints, is the Black-Scholes options-trading model:

Variables
  - *OP*, *SP*, *StrP*, *RFR*, *MT*, *SV*, *D1*, *D2* (as above)
  - Cumulative normal distribution of *D1*: *N1*
  - Cumulative normal distribution of *D2*: *N2*

Constraints
  - $OP = 4.1$
  - $StrP = 40.0$
  - $RFR = 0.1$
  - $SP = 0.2$
  - $0.1 \leq MT \leq 1.0$
  - $D1 = (ln(SP/StrP) + (RFR + 0.5 \times SV^2) \times MT)/SV \times sqrt(MT)$
  - $D2 = D1 - SV \times sqrt(MT)$
  - $N1 = $ cum_normal_distribution($D1$)
  - $N2 = $ cum_normal_distribution($D2$)
  - $OP = SP \times N1 - exp(-RFR \times MT) \times StrP \times N2$

The variables *OP*, *StrP*, *RFR* and *SP* are given a specific value in this invocation of the model. The reason they are not specified as parameters, is that the model can also be invoked with any of these being variables, and other variables being assigned a value instead.

The cumulative probability distribution has a mathematical definition which is assumed to be available (as a "function"). In Chap. 8 we give the Black-Scholes model in MiniZinc, including the specification of the cumulative normal distribution.

Nonlinearity Surprising Results

Nonlinearity can be unintuitive. If you make a journey from location *A* to location *B* and back; on the way from *A* to *B* you travel at 40 km/h and on the way back from *B* to *A* you travel at 60 km/h, what is your average speed? Intuitively this is 50 km/h. But in fact your average speed is distance over *time*, and the time you spend in each direction itself depends on the speed. Specifically your average is

$$\frac{2d}{\frac{d}{40} + \frac{d}{60}} = 48$$

Indeed if you travel one way at 1 km/h and the other way at 100 km/h, your average speed is about 2 km/h!

It is challenging to understand or predict the behaviour of a function of many variables, especially when they are multiplied, or combined in some other nonlinear fashion.

6.7 Summary

This chapter introduces the constraint classes *Finite domain*, *Propositional*, *Linear*, *Integer-Linear* and *Non-linear*. The finite domain class has a finite number of candidate solutions. The domains of variables in propositional constraints have domains comprising just *true* and *false*, and their expressions can be evaluated using truth tables. Linear constraints involve floating point numbers, which cannot practically be represented using finite domains. However the restriction to linear terms, expressions and constraints makes constraints of this class easy to handle. Nonlinear expressions include polynomials and trigonometrical functions, like *sin*. Small problem instances can be solved, as illustrated by the Black-Scholes equation.

Chapter 7
Constraint Classes and Solvers

This chapter explores the complexity of problems involving the different classes of constraints introduced in the last chapter. Each section covers a different class, in order of increasing complexity. As a result the classes are presented in a different sequence from the last chapter.

The constraint classes that will be explored in this chapter are:

- Linear Programming Class
- Propositional Class
- Finite Integer Class
- Mixed Integer-Linear Class
- Non-linear Class

7.1 Linear Programming Models

7.1.1 Linear Expressions and Constraints

In Sect. 6.4.3 the class of linear models was introduced using three definitions: *linear term*, *linear expression* and *linear constraint*.

A *linear term* is either a constant C, or a variable V, or the product of a constant and a variable $C * V$.

A *linear expression* is a sum of linear terms

A *linear constraint* is an equation $Expr1 = Expr2$, or inequation $Expr1 \geq Expr2$ or $Expr1 \leq Expr2$, between two linear expressions $Expr1$ and $Expr2$.

A linear model admits only floating point variables, and no logical connectives.

© Springer Nature Switzerland AG 2020
M. Wallace, *Building Decision Support Systems*,
https://doi.org/10.1007/978-3-030-41732-1_7

7.1.2   Solving Linear Problems

Methods for solving linear problems have a long history, going back to Fourier's method for removing a variable from a set of linear inequations. The simplex method [15], the basis for many modern linear programming solvers, dates from 1947. The method is not simple, but it is quite intuitive.

A linear constraint can be represented as a line in two-dimensional space, as a plane in three-dimensional space and, in general, as an $N-1$ dimensional surface in N-dimensional space. Such a constraint divides the N-dimensional space into two regions: the feasible region on one side and the infeasible region on the other. Note that the solution space for a problem in N variables is an N-dimensional space. A linear constraint in such a problem is represented by an $N-1$ dimensional surface, even if it mentions fewer than N variables. For example the constraint $X > 0$ in a 3-dimensional problem (a problem with three variables, X, Y, Z), is represented by the surface $X = 0$ which is a 2-dimensional space (i.e. the plane $X = 0, Y \in -\infty..\infty, Z \in -\infty..\infty$). An example in three-dimensional space is illustrated in Fig. 7.1.

The feasible region is represented as a solid with flat faces and straight edges. In three dimensions this is termed a *polyhedron* and in N-dimensional space, a *polytope*.

Each vertex is defined by the intersection of three constraints. For example the vertex $X_1 = 0, X_2 = 0, X_3 = 0$ is at the intersection of the three constraints $X_1 \geq 0, X_2 \geq 0, X_3 \geq 0$. In an N-dimensional space a vertex is defined by the intersection of N independent (linear) constraints.

If two points lie on the same side of a linear constraint, then all points on the line joining the two points also lie on the same side. The consequence is that a line between any two feasible points lies entirely within the feasible region. Moreover,

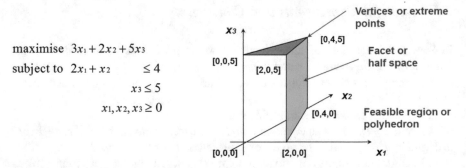

maximise $3x_1 + 2x_2 + 5x_3$

subject to $2x_1 + x_2 \leq 4$

$x_3 \leq 5$

$x_1, x_2, x_3 \geq 0$

Fig. 7.1   A linear problem with three variables

if the optimisation expression is also linear, then on any line the value of the optimisation expression increases in one direction along the line (or else the value of the optimisation expression is the same for all points on the line). It follows that the optimal feasible solution either lies on the surface defined by one or more constraints, or else the optimal value is infinite.

On any line within the surface defined by a constraint, the optimisation expression increases in one direction. Furthermore the intersection of any number of constraints up to $N - 1$ includes a line, and the optimisation function increases in one direction along this line. We conclude that an optimal feasible solution is either infinite, or it lies on the intersection of N independent constraints, which is a single point. Specifically it is a vertex of the polytope.

To find an optimal feasible solution to a set of linear constraints over N variables, it therefore suffices to check all the (finite number of) vertices where N constraints intersect.

The *simplex* method finds the optimal vertex starting from any other vertex of the polytope. The idea is to repeatedly move to a neighbouring vertex at which the optimisation expression takes a higher value. If the current vertex is not optimal, at least one such neighbouring vertex must exist (since the line from the current vertex to an optimal vertex lies inside the polytope). An optimal vertex is reached after a finite number of such steps, and it is proven optimal if no neighbouring vertex has a higher value.

A move from one vertex to a neighbouring one using the simplex method is termed a *pivot*. A sequence of pivots from the vertex $X_1 = 0$, $X_2 = 0$, $X_3 = 0$ to the optimal vertex in our simple example is shown in Fig. 7.2.

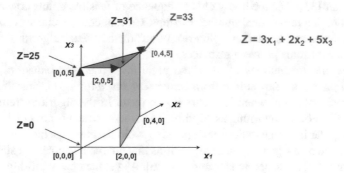

Fig. 7.2 A linear problem with three variables

In our example it is easy to choose the first feasible solution at the vertex where $X = 0$, $Y = 0$, $Z = 0$. In general, there is a way of finding a first feasible vertex for *any* linear program, by solving another linear problem, which is called the *Phase 1 problem*. The Phase 1 problem is created from the original problem by adding an *artificial variable* for each constraint of the original problem.

Specifically for each constraint $Expr_1 \leq Expr_2$ an artificial variable V is added in the Phase 1 problem, with two constraints:

$$V \geq 0$$

$$Expr_1 \leq Expr_2 + V$$

When all the variables in the original problem take the value 0, both these constraints are satisfied by setting V to be the value of $Expr_1 - Expr_2$ if this is positive, and otherwise setting $V = 0$. Choosing this value for each artificial variable V, gives a feasible vertex of the Phase 1 problem. Minimising the sum of the artificial variables gives a feasible vertex of the original problem, if the optimum solution is 0. Otherwise (if the optimum is positive) the original problem is unsatisfiable.

The number of vertices of a polytope can grow exponentially with the number of dimensions in the space. Moreover even moving to better neighbours, the simplex may still visit an exponential number of vertices on the way to an optimum. Consequently the simplex algorithm is—in the worst case—exponential in the number of decision variables [34], though in the "average" case it is polynomial [10].

In 1984 a polynomial algorithm for solving sets of linear equations was published by Kamarkar [33]. The method yields a sequence of feasible points converging on the optimum in a polynomial number of steps. Current commercial solvers support both the simplex and Kamarkar's "interior point" method. Each algorithm has better performance on some problem instances.

Current implementations of all these algorithms used for industrial decision support applications still suffer from *numerical instability*. This instability only arises occasionally, but when it occurs it can result in the algorithm returning an infeasible solution, or returning as "optimal" a solution which is not in fact optimal. Modern popular implementations take pains to avoid the consequences of numerical instability, and a new generation of algorithms that guarantee numerical stability are now emerging from academic laboratories [26]. As yet the new algorithms have not achieved the scalability of the commercial implementations.

Commercial and open-source solvers are available which can find optimal solutions to problems with hundreds of thousands of variables and constraints [53]. Less scalable linear solvers are even available via Excel spreadsheets.

7.2   Propositional Models

7.2.1   *Proposition Expressions and Propositional Clauses*

Propositional expressions have only boolean constants and variables, and logical connectives. The propositional constraint class is called the *SAT* class, and was introduced in Sect. 5.6.2 above as the first example of an *NP*-complete class of problems. A subclass of propositional constraints has been identified as suitable for efficient solving. This is the class of propositional *clauses*. This subclass is expressive enough to represent all propositional constraints.

We introduce two new concepts.

Definition A *literal* is either a propositional constant (*true* or *false*) or a propositional variable (for example X), or a negated variable (for example not X). □

Definition A *disjunction* is a boolean expression of the form
Expr1 or *Expr2* or ... or *ExprN*
where *Expr1, Expr2 ... ExprN* are boolean expressions □

With these two concepts it is simple to define a *clause*

Definition A (propositional) *clause* is a disjunction of literals □

Problems modelled using only propositional clauses are called *k-SAT* problems, where the clauses all have length less than (or equal to) k.

An example of a 3-SAT model is the following, which says that two of the boolean variables X, Y, Z must be true, and one must be false:

$$X \text{ or } Y$$

$$X \text{ or } Z$$

$$Y \text{ or } Z$$

$$(not\, X) \text{ or } (not\, Y) \text{ or } (not\, Z)$$

Naturally propositional clauses belong to the propositional class of models. Thus the class of problems expressible as clauses is a subclass of SAT. Any problem represented using propositional constraints can be translated to clausal form in polynomial time, as shown below. Hence the class of propositional clauses is *NP*-complete.

7.2.2   *Benchmark Results*

Every year there is a conference on propositional problem solving. The problems are typically expressed using the clause representation introduced earlier.

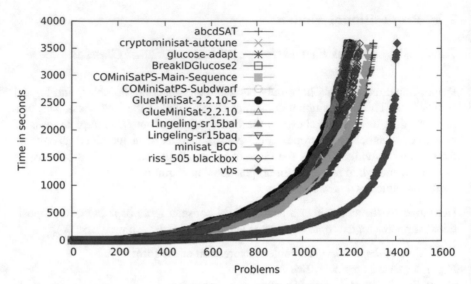

Fig. 7.3 Application benchmark results

The smallest benchmarks problems may have hundreds of variables and thousands of clauses. The largest benchmarks have millions of variables and tens of millions of clauses.

The method of solving propositional problems using truth tables, discussed earlier, has the drawback that for N variables the truth table needs 2^N rows. There is a column for each constraint. Even the smallest benchmarks would therefore need a truth table with more rows than there are atoms in the universe.

3 variables	Truth table has 8 rows
10 variables	Truth table has 1024 rows
100 variables	Truth table has 1,000,000,000,000,000,000,000,000,000,000 rows

However the results of the SAT competition in 2015 [6] on the benchmarks are illustrated in Fig. 7.3. In short, most of the problems were solved within a few hours.

7.2.3 Converting a Propositional Model to Propositional Clauses

To gain the efficiency of SAT solving, in applications such as planning, diagnosis, and computer-aided verification, problem models are often reduced to a representation as a set of clauses [51].

   The process of reduction from a generic propositional model to a set of clauses
has three steps:

1. Replace every expression involving the connectives `implies` and `=`, by an
   expression using only the connectives `not`, `and`, and `or`. Specifically, replace
   Expr1 `implies` *Expr2* by `not` *Expr1* `or` *Expr2*, and replace *Expr1* `=` *Expr2*
   by (`not` *Expr1* `and` `not` *Expr2*) `or` (*Expr1* `and` *Expr2*).
2. Push negations inwards, through `and` and `or`.
   To do this replace each subexpression `not`(*Expr* `and` *Expr*) by (`not` *Expr*) `or`
   (`not` *Expr*), and replace each subexpression `not`(*Expr* `or` *Expr*) by
   (`not` *Expr*) `and` (`not` *Expr*).
3. Push disjunctions inwards through `and`.
   To do this replace each subexpression *Expr* `or` (*Expr1* `and` *Expr2*) by
   (*Expr* `or` *Expr1*) `and` (*Expr* `or` *Expr2*). Similarly replace each subex-
   pression (*Expr1* `and` *Expr2*) `or` *Expr* by (*Expr1* `or` *Expr*) `and` (*Expr2*
   `or` *Expr*).

After each step, simplify the resulting expression. This is not essential but can
dramatically simplify the result.

- drop double negations
- drop the conjunct *true* from any conjunction and the disjunct *false* from any
  disjunction

This process transforms any propositional formula to a SAT formula. For example
the propositional formula

$$(J \text{ or } A) = B$$

is transformed as follows:

1. replace = with propositional connectives
   (*not* (*J or A*) *and not B*) *or* ((*J or A*) *and B*)
2. push negation through disjunction
   ((*not J and not A*) *and not B*) *or* ((*J or A*) *and B*)
3. push disjunction through conjunction
   ((*not J and not A and not B*) *or* (*J or A*)) *and*
   ((*not J and not A and not B*) *or* *B*)
4. push disjunction through conjunction again
   (*not J or* (*J or A*)) *and* (*not A or* (*J or A*)) *and* (*not B or* (*J or A*)) *and*
   (*not J or B*) *and* (*not A or B*) *and* (*not B or B*)

This is now in clause form   but we could drop clauses that are necessarily true
leaving:

$$(not \ B \ or \ J \ or \ A) \ and \ (not \ J \ or \ B) \ and \ (not \ A \ or \ B)$$

This is very different from the original formula (*check it has the same truth table!*), and it is also much longer.

The drawback is that pushing *or* through *and* may, in the worst case, generate a number of clauses which is exponential in the size of the original propositional constraints! The result is a huge SAT model. For example the formula
$(X_1$ *and* $Y_1)$ *or* $(X_2$ *and* $Y_2)$ *or* ... *or* $(X_M$ *and* $Y_M)$
by this transformation yields an exponential number (2^M) of clauses:

$(X_1$ *or* X_2 *or* X_3 *or* ... *or* $X_M)$ *and*

$(Y_1$ *or* X_2 *or* X_3 *or* ... *or* $X_M)$ *and*

$(X_1$ *or* Y_2 *or* X_3 *or* ... *or* $X_M)$ *and*

$(Y_1$ *or* Y_2 *or* X_3 *or* ... *or* $X_M)$ *and*

$(X_1$ *or* X_2 *or* Y_3 *or* ... *or* $X_M)$ *and*

$(Y_1$ *or* X_2 *or* Y_3 *or* ... *or* $X_M)$ *and*

and ... *and*

$(Y_1$ *or* Y_2 *or* Y_3 *or* ... *or* $Y_M)$

There is another transformation which instead of pushing *or* through *and* introduces an additional boolean variable to replace the conjunction. Thus, for example, (*Expr1 and Expr2*) *or Expr* is replaced by the formula *B or Expr*. To express that *B* entails *Expr1 and Expr2* in clausal form we add two clauses:

Expr1 or not B

Expr2 or not B

This second transformation still yields a SAT formula, but this time there is no exponential "blow-up" in the size of the formula. Indeed the above example under the second transformation yields a formula with only $2M$ conjuncts:

X_1 *or not* B_1

Y_1 *or not* B_1

X_2 *or not* B_2

Y_2 *or not* B_2

...

X_M *or not* B_M

Y_M *or not* B_M

B_1 *or* B_2 *or* ... *or* B_M

The beauty of clausal form is that it can represent any propositional constraints by a transformation that is linear in the size of the original constraints, and in particular yields a set of clauses whose total size is also linear in the size of the original propositional model. If the clausal representation can be efficiently solved, then translating a propositional problem to clausal and solving the resulting problem may be an efficient method for solving the original problem. As we saw above, propositional problems with a million variables and a million clauses are routinely solved within minutes of computation time.

We have shown that propositional problems can always be encoded as sets (conjunctions) of *clauses*. Each clause is a disjunction of literals, e.g. *X or not Y or Z*. We note here that a clause with *zero* literals is unsatisfiable, and any problem containing such an empty clause is therefore unsatisfiable.

7.2.4   Resolution for Solving Clausal Models

Any two clauses which contain conflicting literals (such as X and *not* X) can be combined by a process termed *resolution*. When two such clauses are resolved they yield a new clause, called the *resolvent*. The resolvent is simply a clause containing all the literals in the original two clauses, but without the conflicting literals.

For example if we resolve the two clauses

$$X_1 \ or \ not \ X_2 \ or \ X_3 \ or \ not \ X_4$$

$$X_5 \ or \ X_2 \ or \ X_3$$

a new clause is produced

$$X_1 \ or \ X_3 \ or \ not \ X_4 \ or \ X_5$$

Note that if the same literal occurs in both clauses and is not one of the conflicting literals, such as X_3 in the above example, it only appears once in the resolvent. In general, the resolvent is longer than the original two clauses, but in certain cases it can be shorter. For example the resolvent of

$$X_1 \ or \ X_2$$

$$X_1 \ or \ not \ X_2$$

is the clause X_1 which only has one literal.

Resolution is a solving method for propositional problems (expressed in clause form) because, perhaps surprisingly, if a set of clauses is unsatisfiable then repeatedly resolving pairs of clauses will eventually yield the empty clause. If, on the other hand, clauses are repeatedly resolved until no new clauses can be produced, and if the empty clause has not been produced, then the problem is satisfiable.

7.2.5 Encoding Numbers in Propositional Models

There is one major drawback to propositional models, which is their size.

Most practical problems have decisions which take numerical values, such as the number of items that should be purchased in an inventory problem. Such problems are encoded in propositional logic by introducing a propositional variable for each number. If the number of items to be purchased must fall in the range 1..10, then it suffices to introduce an array of 10 propositional variables *ItemCount*[1], ... *ItemCount*[10]. However it the number purchased could be as high as 1000, then 1000 propositional variables are needed.

This translation to propositional form not only leads to a large number of propositional variables, but also a large number of propositional constraints. Suppose there were two types of item, item1 and item2, and for item1 more should be purchased than for item2. Assuming two arrays of decision variables *Item1Count* and *Item2Count*, this constraint would have to be expressed as a constraint on every possible number of items purchased, saying $Item1Count[n] \geq Item2Count[n]$:

not Item1Count$[n] = 1$ *or not Item2Count*$[n] = 2$

not Item1Count$[n] = 1$ *or not Item2Count*$[n] = 3$

...

not Item1Count$[n] = 1$ *or not Item2Count*$[n] = 10$

not Item1Count$[n] = 2$ *or not Item2Count*$[n] = 3$

...

This is the *direct* encoding of integers into booleans. There are alternative encodings, described in [63], which enable different integer constraints to be expressed more compactly.

7.2.6 Lazy Clause Generation

The drawback of propositional modelling is the huge number of clauses required to model a problem—particularly if it includes numbers. Modelling with finite integers, as described in the next section, supports much more compact and readable models.

The technique of Sect. 9.7.2 deploys the technology of propositional solvers while it avoids this explosion in model size, by modelling problems with finite integer models. Propositional clauses are generated dynamically by the algorithm only where they are needed to find a solution. It is further described in Sect. 9.7.2.

This powerful method is available in MiniZinc as the "Chuffed" solver listed in Sect. 8.6 below.

7.3 Finite Integer Models

7.3.1 Finitely Many Combinations to Check

To solve a finite integer model it suffices to try each combination of values for the variables and check them against all the constraints.

On a current state-of-the-art (deterministic) computer, there are much better algorithms than generate-and-test. Most such algorithms depend on some form of search, which will be described in Chap. 9.

However using the table representation of finite integer constraints, it is possible to solve the problem without search, as described in this section.

7.3.2 Tables

The number of alternative complete assignments to the decision variables in a finite integer model is the product of their domain sizes. The challenge in finding feasible and optimal solutions is to avoid having to check all these alternatives. Since the number of complete assignments is finite, this is also true of all partial assignments. In particular for any constraint in a finite integer model, the number of alternative assignments to the variables in its scope is just the product of the sizes of their domains.

Take the assignment problem (from page 23) for example, with 3 agents and 3 tasks, and the constraint $TV[1] = value[A[1], 1]$. We can write down all ways of satisfying this constraint:

$TV[1] = 1, A[1] = 1$
$TV[1] = 4, A[1] = 2$
$TV[1] = 7, A[1] = 3$

The scope of the constraint $TV[1] = value[A[1], 1]$ is the set of variables $\{TV[1], A[1]\}$. These can always be written out as a table as in Table 7.1:

For simply checking the constraint—for example checking that if task 1 is assigned to person 1 and the cost is 1—there is no need to construct this table. It suffices to check that $1 = value[1, 1]$. However the table enables us to efficiently evaluate combinations of constraints.

Table 7.1 Constraint **C1**:
$TV[1] = value[A[1], 1]$

C1	TV[1]	A[1]
	1	1
	4	2
	7	3

Table 7.2 Constraint **C2**:
$TV[2] = value[A[2], 2]$

C2	TV[2]	A[2]
	2	1
	4	2
	6	3

Table 7.3 Constraint **C3**:
$A[1] \neq A[2]$

C3	A[1]	A[2]
	1	2
	1	3
	2	1
	2	3
	3	1
	3	2

Table 7.4 Constraint **C4**: **C1** and **C3**

C4	A[1]	A[2]	TV[1]
	1	2	1
	1	3	1
	2	1	4
	2	3	4
	3	1	7
	3	2	7

Table 7.5 Constraint **C5**: **C4** and **C2**

C5	A[1]	A[2]	TV[1]	TV[2]
	1	2	1	4
	1	3	1	6
	2	1	4	2
	2	3	4	6
	3	1	7	2
	3	2	7	4

We can write out the table for two more constraints, **C2** (Table 7.2) and **C3** (Table 7.3). Combining constraints **C1** and **C3**, dropping all the rows that conflict with each other, we can form a new Table 7.4. We call this *joining* the two tables, and the result is **C4**, the *join* of tables **C1** and **C3**. We can then join the tables **C4** and **C2**, via Table 7.5. Next add the Table 7.6 for $\sum_{i \in 1..3} TV[i] \geq 14$: and join these tables together, leaving the following wider Table 7.7: Gradually the set of all feasible assignments to the problem variables can be constructed by joining the tables from all the constraints.

Table 7.6 Constraint **C6**:
$\sum_{i \in 1..3} TV[i] \geq 14$

C6	TV[1]	TV[2]	TV[3]
	4	6	4
	4	6	5
	7	4	3
	7	4	4
	7	4	5
	7	6	3
	7	6	4
	7	6	5

Table 7.7 Constraint **C5** and **C6**

C5 and C6	A[1]	A[2]	TV[1]	TV[2]	TV[3]
	2	3	4	6	4
	2	3	4	6	5
	3	2	7	4	3
	3	2	7	4	4
	3	2	7	4	5

Table 7.8 Constraint **C7**: **C1** and **C2**

C1 and C2	A[1]	A[2]	TV[1]	TV[2]
	1	1	1	2
	1	2	1	4
	1	3	1	6
	2	1	4	2
	2	2	4	4
	2	3	4	6
	3	1	7	2
	3	2	7	4
	3	3	7	6

7.3.3   Minimising the Number of Rows in Intermediate Tables

The number of rows in the tables produced can be kept as small as possible by combining them in the right sequence. In the assignment example, suppose we first joined the tables **C1** and **C2**, as shown in Table 7.8. The join would have 9 rows: This table has more rows than **C1 and C3**, so is a more costly intermediate when constructing **C1 and C2 and C3**. In such a small example the difference is insignificant, but with industrial applications where a table may have 10,000 rows it makes a huge difference.

The number of rows in a constraint with two variables cannot be larger than the product of their domains. If the number of rows equals this product, then every value in the domain of each variable appears in a row with every value in the domain of the other variable. Such a constraint is automatically satisfied by every assignment, so it has no effect on the solutions to any model in which it occurs.

The fewer rows there are in the table for a constraint the more effect it could have in reducing the set of feasible solutions. We call a constraint with fewer rows a *tighter* constraint.

To keep the size of intermediate tables as small as possible the tables to join first are:

1. tighter constraints
2. constraints whose scopes have a bigger overlap (i.e. more variables in common)

These are useful heuristics (hints) for choosing the order in which to join the constraint tables. Unfortunately there is no easy way to find an order guaranteed to yield the smallest intermediate constraints.

7.3.4 Minimising the Number of Columns

For a fixed size D of variable domains, the number of rows in a table can be no greater than D^W where W is the *width* of the table—i.e. the number of columns. By ensuring that intermediate tables have a maximum width of W, we can therefore ensure that the intermediate tables have less than D^W rows.

If there are many decision variables, then the final table must have a large width. However for many problems, though the problem may involve many variables, the user may only be interested in whether the problem is satisfiable, and not the values of all the variables in each solution. (Similarly for optimisation problems the user may only be interested in the optimal value and not in particular optimal solutions.) In these cases we can choose the order of joins so as to minimise the width of intermediate solutions.

The advantage of minimising the table width is that the limit on the number of rows in intermediate tables is not dependent on the particular problem instance. More precisely it does not depend on the actual rows that appear in the tables representing the constraints. Indeed if an order of joins can be found for which no intermediate table has width greater than a constant K, then the worst case complexity of the solver is polynomial in D^K, where D is the domain size., i.e. polynomial in D.

The mechanism that a solver can use to keep the width of intermediate tables small is to *project* a table onto the variables of interest, dropping the columns for the remaining variables and yielding a smaller table. In particular, once the additional columns have been dropped, there may be many identical rows, enabling rows to be dropped as well, until there is just one representative of each set of identical rows.

The Constraint Graph
The choice of the join order that minimises the maximum width of any intermediate table is best illustrated by means of a graph. The *constraint graph* of a problem is a graph in which each node represents a problem variable, and each edge represents a constraint. Since a constraint may involve more than two variables, the edge in our

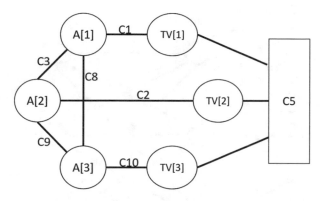

Fig. 7.4 The constraint graph

constraint graph may have edges which connected more than two nodes. Such edges are termed *hyper-edges* and a graph containing hyper-edges is termed a *hyper-graph* [36, Section 1.2]. The hypergraph for our running example with six variables
$A[1]$, $A[2]$, $A[3]$, $TV[1]$, $TV[2]$, $TV[3]$
and 7 constraints
C1,C2,C3,C5, **C8**: $A[1] \neq A[3]$, **C9**:$A[2] \neq A[3]$, **C10**: $value[A[3], 3] = TV[3]$
is shown in Fig. 7.4.

Assuming the model is only used to find a feasible solution, and not the choices of decision variable assignments that lead to it, there are sequences of joins and projections, *with all intermediate tables having 3 or fewer columns*. Note that for this purpose join and projection are distinct operations, each generating an intermediate table.

Try and find such a sequence—there's an answer at the end of Sect. 7.3.

In this example minimising the number of columns in intermediate constraints may not yield the same sequence as minimising the number of rows. The next subsection discusses a class of problems for which projecting out columns is key to their good performance.

The join ordering which minimises table width corresponds to a concept termed *tree-width*, which applies to graphs [32]. A tree decomposition is a mapping of a graph into a tree. The tree-width measures the number of graph vertices mapped onto any tree node in an optimal tree decomposition. The minimal width join ordering corresponds to the tree-width of the underlying constraint graph. Unfortunately finding the tree width of a graph is itself an *NP*-hard problem [5].

7.3.5   *Dynamic Programming*

This section presents a simple optimisation problem as an example of an approach termed *dynamic programming*.

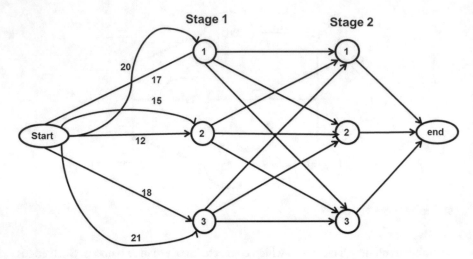

Fig. 7.5 The shortest path problem

The problem is to plan a minimal cost journey from *start* to *end* as illustrated in Fig. 7.5. The figure shows a cost associated with each edge. For each stage we introduce a variable X whose value is the node reached at that stage, and a variable Z representing the cost up to that stage. The problem can be modelled with a constraint for each stage:

- Parameters

    - Locations, *start*, *stage1:1*, . . . *stage2–3*, *end*
    - Lengths of the edges between locations, for example in Table 7.9

- Variables

    - X1 which had domain 1..3
    - Z1 which has domain 12..21
    - X2 . . .
    - Z2 . . .

- Constraints

    - cons1 on the cost of paths from *start* to *stage1*, whose tuples are given in Table 7.9
    - cons2 on the cost of paths from *stage1* to *stage2*
    - cons3 on the cost of paths from *stage2* to *end*

- Objective

    - Minimize Z1+Z2+Z3

We represent *cons1* as Table 7.9. Before joining the table for *cons1* with the table for *cons2*, we can remove three rows from it, which could never be involved in an

Table 7.9 Constraint for shortest
path Stage 1

	X1	Z1
Start	1	20
Start	1	17
Start	2	15
Start	2	12
Start	3	18
Start	3	21

Table 7.10 Reduced
constraint for Stage 1

X1	Z1
1	17
2	12
3	18

optimal solution. Naturally we only need to consider the cheapest edge to each of
the alternative nodes at Stage 1. We can also project out the first column which is
not involved in any other constraints. The reduced table is thus Table 7.10.

After joining the reduced *cons1* and *cons2*, the variable $X1$ can be projected out
of the resultant table, as it does not appear in any further constraints. Note also that
the resultant table can now be reduced by removing all rows ending at the same node
in Stage 2, for which $Z1 + Z2$ is non-minimal. This optimisation is a toy example
of a solving method called *dynamic programming* [17].

7.3.6   Complexity of Finite Integer Models

The constraints we have encountered have simple comparisons such as $=$, $\geq$ and
so on. The expressions on each side of the comparison are constructed using
mathematical functions, such as $+$ and $\times$ or else they require looking up an entry in
an array or a matrix. All the operations are simple: constant time, or logarithmic in
the size of the numbers involved, but never worse than polynomial in the size of the
input. Given a complete assignment of values to the decision variables of a model,
it is therefore relatively easy (i.e. polynomial) to check all the problem constraints
to determine whether they are satisfied.

However the number of possible complete assignments to a set of n variables
$v_1, \ldots v_n$ is $d_1 \times \ldots \times d_n$ where d_i is the size of the domain of the ith variable. This
number is greater than d^n where d is the size of the smallest domain. Assuming
$d \geq 2$, the number of complete assignments grows exponentially with the number
of variables.

If the algorithm tries the different alternative values in the domain of each variable, this corresponds to the generate and test algorithm introduced in Sect. 5.2. Assuming the constraint checking takes polynomial time, as discussed in Sect. 5.5, the resulting algorithm is polynomial, and since this algorithm works for any model in the finite integer class, the class has complexity *NP*.

We have shown above a simple encoding of boolean variables into zero-one variables, and of propositional connectives into integer constraints on zero-one variables. It follows that propositional models can be encoded as finite integer models, and since the class of propositional models is *NP*-complete, it follows that finite integer models are *NP*-hard.

Indeed we can quickly show that any clausal problem can be encoded using finite integer domains. Each propositional variable can be modelled as a numeric variable with domain 0, 1. The negation of a variable X can be encoded as $1 - X$, and a clause with n literals, $lit_1 \ or \ \ldots \ or \ lit_n$, can be encoded as a constraint $lit_1 + \ldots + lit_n \geq 1$. This completes the finite integer encoding. Accordingly we say that SAT is *reducible* to the finite integer class, and therefore the finite integer class is *NP*-hard.

On the other hand finite integer models that are decision problems rather than optimisation problems, are in the class *NP*. It follows that the class of finite integer decision models is both in *NP* and *NP*-hard so it is *NP*-complete.

7.3.7 Answer to Minimising Intermediate Table Widths

One example sequence of joins and projections is:

1. Join **C1** and **C3** yielding **JC13**
2. Project **JC13** onto $A[2]$ and $TV[1]$ yielding **C13**
3. Join **C13** and **C2** yielding **JC123**
4. Project **JC123** onto $TV[1]$ and $TV[2]$ yielding **C123**
5. Join **C123** and **C5** yielding **JC135**
6. Project **JC135** onto $TV[1]$ and $TV[3]$ yielding **C135**
7. Project **JC135** onto $TV[2]$ and $TV[3]$ yielding **C235**
8. Join **C135** and **C1** yielding **JC15**
9. Project **JC15** onto $A[1]$ and $TV[3]$ yielding **C15**
10. Join **C15** and **C10** yielding **JC110**
11. Project **JC110** onto $A[1]$ and $A[3]$ yielding **C110**
12. Join **C110** and **C8** yielding **CFinal1**
13. Join **C135** and **C2** yielding **JC25**
14. Project **JC25** onto $A[2]$ and $TV[3]$ yielding **C25**
15. Join **C25** and **C10** yielding **JC210**
16. Project **JC210** onto $A[2]$ and $A[3]$ yielding **C210**
17. Join **C210** and **C9** yielding **CFinal2**

If **CFinal1** has one or more rows, and **CFinal2** has one or more rows, then the model is satisfiable (Fig. 7.6).

Fig. 7.6 The constraint
graph

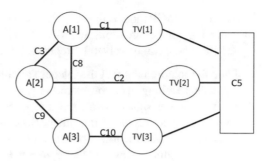

7.4   Linear Constraints over Integers and Floats

7.4.1   Unbounded Integer Constraints

An unbounded integer constraint has within its scope one or more integer variables
that have no bounds. A simple example is the constraint $A^N + B^N = C^N$ where
$A, B, C > 0$ and $N > 2$, which Fermat conjectured to be unsatisfiable. Clearly this
cannot be captured in one or more finite tables, and we have no process to establish
whether or not it is satisfiable.

The reason for requiring that integer expressions are finite in this class of
constraints is to ensure that models involving finite integer constraints can be solved
automatically.

Indeed a solver that could handle unbounded integer constraints was proven not
to be possible by Gödel in 1931 (see [44]). Gödel introduced a way of numbering
every logical formula, and even every sequence of formulae that corresponds to a
logical proof. Gödel's proof used this numbering to write formulae such as $p(N, M)$
which is true if the number N encodes a proof of the formula numbered M.

Gödel's proof can be summarised as follows:

- There is an (unbounded) integer formula (which is encoded by the number M)
- This formula expresses that there exists no integer which encodes a proof of M
- If it was provable it would be false, which is impossible (assuming the proof
  system is consistent)
- It must be unprovable, and (therefore) true
- Therefore the formula encoded by M is true, but cannot be proven
- A solver that showed M is satisfiable would thereby create a proof of M
- Therefore no solver can be complete for unbounded integer arithmetic

7.4.2   Branch and Bound

The "mixed integer-linear programming" (MIP) class comprises problems that are
modelled using only linear expressions over floating point and integer variables.

The class of MIP models therefore looks very similar to the class of linear models over floating point expressions, but MIP models are in general much harder to solve.

A MIP model can be solved by repeating two steps.

1. Drop the requirement that the integer variables can only take integer values, and solve the resulting linear problem.
2. If the values assigned to all the integer variables are integers, then this is a solution. Otherwise take a non-integer assignment to an integer variable (such as $X = 1.5$) and add the lower bound constraint ($X \geq 2$) that excludes that integer solution, creating a new problem **Problem1**. Find solutions to **Problem1** starting at step 1. For completeness it is also necessary to add the alternative upper bound constraint ($X \leq 1$), creating another problem **Problem2**, and finding solutions to this problem.

There are two alternative subproblems generated for each integer variable which gets assigned a non-integer value in the solution of the previous relaxed problem. In the "worst case" this may occur (multiple times) for each integer variables in the original MIP. However it is provable that the number times any variable can take a non-integer value that has not been previously excluded is finite.

In case the optimal solution is sought, the search can be adapted to avoid having to find all the solutions to the original MIP. The approach relies on the fact that the optimal solution to the linear relaxation of a problem is never worse than the optimal solution of the original MIP problem.

In the enhanced algorithm, once any solution to the MIP problem has been found it is recorded as the current optimum (which we can call *curr-opt*). Now the above algorithm can be improved as follows:

1. Drop the requirement that the integer variables can only take integer values, and solve the resulting linear problem. If its optimal solution is better than *curr-opt*, then continue to step 2.
2. If the values assigned to all the integer variables are integers, then record this as the new optimal solution in *curr-opt*. Otherwise take a non-integer assignment to an integer variable and add a lower bound constraint that excludes that integer solution, creating a new "subproblem" **Problem1**. Find solutions to **Problem1** starting at step 1. Also, add the alternative upper bound constraint, creating another subproblem **Problem2**, and finding solutions to this problem.

If different subproblems are sent to different computers, when one computer updates *curr-opt*, this new information should be shared with the other computers. The optimal value is the final value of *curr-opt* when all the alternative subproblems have been solved. The approach is called *branch and bound* [35] and will be further described in Sect. 9.6.3 and works surprisingly well!

7.4.3   Cutting Planes

There is another way to solve MIP problems, which also ultimately relies on solving a linear problem. In this case, however, the new constraints that are added are all entailed by the original MIP problem, so there is no need to explore alternatives.

The idea can be illustrated on a constraint $2 \times X + 2 \times Y \leq 3.5$ with two integer variables, X and Y. Firstly since X and Y must take integer values any any feasible solution, $2 \times X + 2 \times Y$ must also be integer. Thus if $2 \times X + 2 \times Y \leq 3.5$ it must also be true that $2 \times X + 2 \times Y \leq 3$. However we can improve the constraint still further, by noting that $2 \times X + 2 \times Y = 2 \times (X + Y)$ will always be an even number, so in any feasible solution $2 \times X + 2 \times Y \leq 2$. This constraint can be simplified to the equivalent constraint $X + Y \leq 1$.

By combining constraints it is possible to add more and more constraints, until the optimal solution to the extended set of constraints is guaranteed to be integer-valued [27]. (Indeed it can be guaranteed that any vertex of the polytope of the linear relaxation is integer-valued). Thus the linear solver can be used to solve the original MIP problem. The additional constraints added when solving an integer linear problem are termed *cuts*.

In the context of mixed integer programs which include both integer variables and floating point variables it is still possible by adding a finite number of cuts, all entailed by the original MIP model, to generate a model whose linear relaxation satisfies the integrality of all its integer variables [16]. Nevertheless the number of cuts generated can be exponential.

7.4.4   Logical Combinations of Linear Constraints

Often the requirement for MIP models is in order to handle logical combinations of linear constraints. In this section we describe an approach which can translate logical combinations of linear constraints, whose expressions are bounded by some finite maximum and minimum values, into a set of MIP constraints.

The *reification* of a constraint is simply a boolean variable which is logically equivalent to the constraint.

Definition Let C be an arbitrary constraint involving one or more variables. A boolean variable B is a *reification* of the constraint C if $C = B$. If the constraint C is satisfied, $B = true$ and if C is unsatisfied then $B = false$.   □

For bounded floating point constraints we introduce a closely related concept which we call *integer half-reification*. Half-reification is a weaker version of reification which catches the implication in one direction. Its definition is similar:

Definition Let C be an arbitrary constraint involving one or more variables. A variable B with domain $0, 1$ is an *integer half-reification* of the constraint C if $(B = 1)$ implies C. If $B = 1$ then C is satisfied, and if C is unsatisfied then $B = 0$.   □

The method for handling logical combinations of linear constraints, is to formulate an integer half-reification of each constraint and then constrain the introduced 0, 1 variables. If $B1$ and $B2$ are integer half-reifications of two constraints $C1$ and $C2$ then $C1$ *or* $C2$ is enforced by the constraint $B1 + B2 \geq 1$.

Consider the finite integer constraint $Expr1 \geq Expr2$. Because the expressions are bounded, there is a maximum value $Max2$ that can be taken by the expression $Expr2$, and there is a minimum value $Min1$ that can be taken by the expression $Expr1$. We introduce a variable B with domain 0, 1. The following constraint ensures that B is the integer half-reification of $Expr1 \geq Expr2$:

$Expr1 + (1 - B) \times (Max2 - Min1) \geq Expr2$

A concrete example will illustrate this more clearly. The constraint below ensures that B is the integer half-reification of $X > Y$.

Parameters – $MinX = 1$
            – $MaxY = 6$
Variables   – X with domain 1..5
            – Y with domain 3..6
            – B with domain 0..1
Constraints – $X + (1 - B) \times (MaxY - MinX) \geq Y$

Firstly if $X = 1$ and $Y = 6$, then the constraint is $1 + (1 - B) \times 5 \geq 6$, which is satisfied only if $B = 0$. Indeed if $X < Y$ then $B = 0$. However if $X = 5$ and $Y = 3$, then the constraint is $5 + (1 - B) \times 5 \geq 3$, which is satisfied both by $B = 0$ and by $B = 1$. Indeed if $X \geq Y$ then the constraint is satisfied by both $B = 0$ and $B = 1$.

The value *Max2-Min1* is designed to be large enough so that the constraint itself never fails. The boolean variable B is forced to take the value 0 if $X < Y$, but once $B = 0$, it does not matter what values are taken by the variables X and Y. Even if X takes its smallest possible value and Y its largest, the constraint is still satisfies when $B = 0$. Any number large enough, like *Max2-Min1* to have this property is termed a *BigM* number. The constraint

$$X + (1 - B) \times BigM \geq Y$$

where *BigM* is a large enough constant so that the constraint never fails, is often called a *BigM constraint*.

As a toy example this finite integer model illustrates the handling of disjunction using half-reification.

Variables   X with domain 0..5
Constraints $X = 0$ *or* $X \geq 3$

is translated to the following finite integer model:

Variables – X with domain 0..5
           – B1 with domain 0..1
           – B2 with domain 0..1

Constraints $- (1 - B1) \times 5 \geq X$
$ - X + (1 - B2) \times 3 \geq 3$
$ - B1 + B2 \geq 1$

7.4.5 Linear Constraints over Integer Variables

The class of finite integer constraints and the class of linear constraints over integer variables have a common subclass. These are linear constraints over finite integer variables. Constraints in this class can be solved either as finite integer constraints (for example using tables, joins and projections) or as linear constraints (for example using branch and bound).

Since branch and bound is often an efficient method of solving problems, it is very useful to convert logical combinations of linear constraints into just linear constraints, thus linearising a finite integer model.

Instead of half-reification, we can use reification in case the relevant variables are integer. If the variables X and Y are integer variables, we can also enforce the implication in the other direction: that $X \geq Y$ implies $B = 1$. As before, we introduce $MaxX$, the maximum possible value of X and $MinY$ the minimum of Y. The required constraint is

$$Y + B \times (MaxX + 1 - MinY) \geq X + 1$$

Now if $X = 5$ and $Y = 3$ we have $3 + B \times 3 \geq 6$, which enforces $B = 1$, and similarly for any $X \geq Y$.

In case the variables have integer domains, we can handle all logical combinations of constraints using a full reification:

$$(B = 1) \leftrightarrow (X \geq Y)$$

and

$$(B = 0) \leftrightarrow (Y \geq X + 1)$$

Now logical combinations of constraints are simply expressed as finite integer constraints on their 0, 1 variables. The translations of the propositional connectives into constraints on their reified boolean variables is given in the Table 7.11.

Table 7.11 Zero-one implementation of propositional connectives

Not C1	C1 and C2	C1 or C2	C1 implies C2	C1 = C2
B1=0	B1+B2 = 2	B1+B2 ≥ 1	B1 ≥ B2	B1 = B2

In the light of these results, propositional combinations of constraints can be used in finite integer models.

7.4.6 Complexity of Linear Constraints over Integers and Floats

The Integer Linear constraint class is *NP*-hard. It contains an *NP*-complete class as a subclass, but it also admits optimisation, which takes it outside the class *NP*.

7.5 Nonlinear Constraints

Nonlinear constraints are all those constraints that can't be efficiently modelled using linear expressions over floats. Integer-linear constraints enable us to model logical combinations of linear constraints. Finite integer constraints allow us to model non-linear expressions over finite integer variables.

The nonlinear constraints discussed in this section are the ones that are not efficiently modelled in other constraint classes. These involve nonlinear expressions over real numbers, modelled as float constants and variables.

7.5.1 Solving Non-linear Problems

The simplest scenario for non-linear problem-solving is an equation involving a single variable. Any equation *Expr1* = *Expr2* can be written in the form *Expr1* − *Expr2* = 0, so to satisfy the constraint it is necessary to find a zero of the function *Expr1* − *Expr2*.

Let us assume we have found two values for X, a "minimum" value s for which the function takes a negative value, and a "maximum" value l for which the function is positive. Our solving method depends upon the function being continuous in the interval $s..l$. Clearly under these assumptions there is a value for X in the interval $s..l$ where *Expr1* − *Expr2* = 0. The requirement for the solver is to find this value.

It does this by testing the value *fmid* of *Expr1* − *Expr2* at the mid point of the interval $X = \frac{s+l}{2}$. If *fmid* = 0 then the constraint is satisfied. If *fmid* < 0 then we can reduce the interval to *fmid..l*, and find the zero within this smaller interval. If *fmid* > 0 then we can reduce the interval to *s..fmid*, and find the zero within this smaller interval. This method is termed *binary approximation*, and is illustrated in Fig. 7.7. Assuming a binary representation of the value of X, we find the solution to one more significant figure each time we halve the width of the interval.

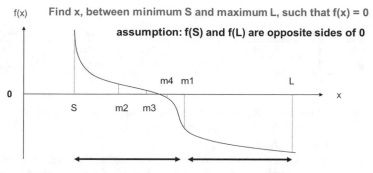

Find x, between minimum S and maximum L, such that f(x) = 0

assumption: f(S) and f(L) are opposite sides of 0

After n bisections, the width of the interval is $(L-S) / 2^n$
- every three bisections the answer is accurate to another significant figure

Fig. 7.7 Binary approximation for continuous functions

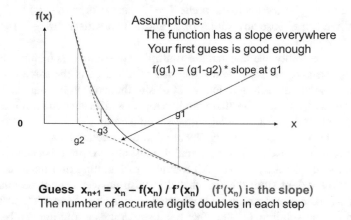

Assumptions:
The function has a slope everywhere
Your first guess is good enough

$f(g1) = (g1-g2) * \text{slope at } g1$

Guess $x_{n+1} = x_n - f(x_n) / f'(x_n)$ ($f'(x_n)$ is the slope)
The number of accurate digits doubles in each step

Fig. 7.8 Newton's method for differentiable functions

If there are more than one variables in *Expr1 − Expr2* the same procedure solves the constraint, but now each variable has an interval, and each interval must be halved in turn until the required precision has been achieved.

The rate of improvement can be improved dramatically if the expression *Expr1 − Expr2* has a slope at each point. In Fig. 7.8 we write $f(X)$ for the function to be zeroed by the constraint, instead of *Expr1 − Expr2*.

Instead of starting with two points, one on each side of the solution, we need only a single starting point, here called $g1$. Assuming $g1$ is close enough to the point where f takes the value 0, our next guess is the point $g2$ which is reached by "sliding down the slope" from $g1$. The next guess, $g3$, is in turn reached by sliding down the slope at $g2$. This is called *Newton's method*.[1]

---

[1] Also called the Newton-Raphson method.

Table 7.12 Accuracy of Newton's approximation

X	10	35.6	26.396	24.790	24.73869	24.7386338
$X^2 - 612$	-512	655.36	84.75	2.54	0.003	0.0000023
Accurate digits			2	24.7	24.7386	24.7386338

As a simple illustration of its convergence, we apply this method to solve the constraint $X^2 - 612 = 0$, given a starting point of $X = 10$. The results at each successive iteration are given in Table 7.12.

Now, suppose we have multiple variables in a constraint, then we can slide down in many different directions. For example suppose there are two variables, the first guess is $X = x_0$ and $Y = y_0$, and $f(x_0, y_0) = c_0$. Suppose the slope in the X direction is $f'x_0$ and the slope in the Y direction is $f'y_0$. Then we can slide down in the X-direction to the point where $(x_0 - X) \times f'x_0 = c_0$ and $Y = y_0$ or we could slide down in the Y-direction to the point $X = x_0$ and where $(y_0 - Y) \times f'y_0 = c_0$. In general we can choose any values of X and Y that satisfy $(x_0 - X) \times f'x_0 + (y_0 - Y) \times f'y_0 = c_0$.

With this flexibility we can handle multiple constraints, and more and more closely approximate the point where they are all satisfied at the same time. To do this we must solve, at each iteration, a set of simultaneous equations. Specifically, we find values of the variables that simultaneously satisfy an equation like the one above for the function f associated with each constraint.

Two drawbacks illustrated in Fig. 7.9 are that the functions may not have a slope everywhere, that the Newton approximation may not converge because the initial point was too far from a solution. All methods of handling non-linear constraints suffer to some extent from the fact that a precise answer cannot be guaranteed. Consequently sometimes a solution is returned (that is close "enough") even though there is in fact no solution. On the other hand sometimes a solution is missed, even when there is one. Thirdly sometimes two solutions are so close together that they are not told apart by the solver.

Nevertheless, we have broadly a method for approximating, as closely as needed, a point in space that simultaneously satisfies multiple non-linear constraints.

Modern techniques for solving nonlinear problems even when they don't have a slope use interval reasoning, and have been implemented in systems such as Numerica [64] and Realpaver [28]. The slope can also be used in interval reasoning to improve its search behaviour.

7.5.2   Complexity of Nonlinear Constraints

There is no practical reason to encode propositional constraints as nonlinear constraints on float variables, but the fact that it is possible tells us about the expressive power of nonlinear constraints, and has consequences for their complexity.

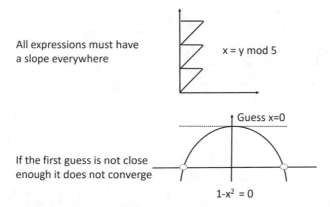

All expressions must have
a slope everywhere

x = y mod 5

Guess x=0

If the first guess is not close
enough it does not converge

$1-x^2 = 0$

Fig. 7.9 Drawbacks of Newton's method

In short the constraint $x \times (1 - x) = 0$ on the float variable x is satisfied if $x = 0$ or $x = 1$. This constraint enables a float variable to be constrained to be a zero-one variable. All the propositional connectives can now be expressed as nonlinear expressions, as was done previously when representing propositional expressions as finite integer expressions.

Since propositional constraints can all be expressed as nonlinear constraints over float variables, the nonlinear constraint class is *NP*-hard. It contains an *NP*-complete class as a subclass, but it also admits nonlinear constraints which allow models to be expressed which do not belong to the class *NP*.

7.6   Summary

This chapter has presented methods for solving different kinds of problems. For linear problems, whose feasible region is described by a polytope with a finite number of vertices, the simplex method "pivots" from vertex to (improving) vertex until an optimum is found. Propositional problems in the class SAT, are transformed to sets of propositional clauses, and solved using resolution as an underlying reduction method. Finite integer models can be represented using tables which can be handled by the join and projection operations widely used in relational databases. This method can be used to implement dynamic programming, for example. For integer linear problems, branch-and-bound and cutting planes are deployed, and reification is used to handle logical constraints.

Finally non-linear problems are traditionally tackled using Newton's methods, with interval reasoning as both an alternative and an extension.

Chapter 8
Constraint Classes in MiniZinc

In this chapter the different classes of models will be modelled using MiniZinc. Three additional features of MiniZinc will be introduced, *logical connectives*, *comprehensions*, *predicates* and *global constraints*. These features enable models to be built from components, and can enable specialised solving methods to be applied to handle certain constraints.

8.1 Logical Connectives

In Sect. 6.2.2 the logical connectives *and*, *or*, *not* and *implies* were introduced. These have a direct representation as operators in MiniZinc. The syntax looks a bit odd, but is inherited from a long mathematical tradition:

and is represented in MiniZinc as /\
or is represented in Minizinc as \/
not is represented in Minizinc as not
implies is represented in Minizinc as ->

Thus in the boats and docks problem (Sect. 2.2.3) the constraint that "Either boat b1 must go in dock4 or b3 must go in dock1" is expressed in MiniZinc as:

```
Boat1 = dock4 \/ Boat3 = dock1
```

and "It is not possible for both b1 and b2 to go in dock4" is:

```
not (Boat1=dock4 /\ Boat2=dock4)
```

The boolean politics example of Sect. 6.3 with the two statements:

Julie	Exactly one of us is telling the truth
Anthony	At least one of us is telling the truth

© Springer Nature Switzerland AG 2020
M. Wallace, *Building Decision Support Systems*,
https://doi.org/10.1007/978-3-030-41732-1_8

is encoded in MiniZinc as:

```
var bool:J ;
var bool: A ;

constraint J = ((J \/ A) /\ not (J /\ A)) ;
constraint A = (J \/ A) ;

solve satisfy ;
```

To write a long conjunction, we can use the `forall` construct.

```
constraint X[1]=1 /\ X[2]=1 /\ X[3]=1 /\ X[4]=1 /\ X[5]=1 ;
```

can be written as:

```
constraint forall(i in 1..5)(X[i]=1) ;
```

Similarly there is a way of expressing a long disjunction using `exists`.

```
constraint X[1]=1 \/ X[2]=1 \/ X[3]=1 \/ X[4]=1 \/ X[5]=1 ;
```

can be written as:

```
constraint exists(i in 1..5)(X[i]=1) ;
```

This can be illustrated by extending the assignment problem (of Sect. 4.3.2) include the following requirement: *There must be one task assigned to the agent who brings the most value to that task.*

```
enum people ;
enum tasks ;
array [people,tasks] of int: value ;
set of int: vals;
int: limit ;
include "assignment_data.dzn" ;
array [tasks] of var people: A ;
array [tasks] of var vals: TV ;
var int: TotalValue ;
constraint
    forall(t1 in tasks, t2 in tasks where t1 != t2)
    (A[t1] != A[t2]) ;
constraint forall(t in tasks)(TV[t] = value[A[t],t]) ;

% New constraint
constraint
   exists(t in tasks)(TV[t] = max(p in people)(value[p,t])) ;

constraint TotalValue = sum(t in tasks)(TV[t]) ;
constraint TotalValue >= limit ;
solve satisfy ;
```

Although *forall* and *exists* are not usually classified as "propositional" connectives, in MiniZinc they can only range over a finite domain of alternatives, so they are propositional in the sense that they always correspond to a set of conjunctions and a set of disjunctions respectively.

8.2 Comprehensions

In previous chapters, sets and arrays have been given explicitly, by listing their elements. For example the array *items* in the knapsack problem was initialised by

```
items = [7,10,23,13,4,16] ;
```

and the set *vals* in the assignment models was initialised by:

```
vals = {18,13,16,12,20,15,19,10,25,19,18,15,16,9,12,8} ;
```

Sometimes an expression can be used to initialise all the elements in an array or a set. The simplest example is when a model requires an array of increasing numbers, for example an array $incs5 = [1, 2, 3, 4, 5]$.

MiniZinc supports this through a "comprehension":

```
array [1..5] of int: incarray =   [j | j in 1..5] ;
```

The same comprehension syntax can be used to define a set:

```
set of int: incset = {j | j in 1..5} ;
```

The form of a comprehension is *expression | generator*. Any expression can be used, for example $2 \times j$ for even numbers, thus:

```
[2 * j | j in 1..5] = [2,4,6,8,10] ;
```

There may be more than one variable in the generator, for example:

```
[j+k | j in 1..2, k in 1..3] = [2,3,4,3,4,5] ;
```

The set comprehension removes repeated elements:

```
{j+k | j in 1..2, k in 1..3} = {2,3,4,5} ;
```

If *value* is the matrix:

```
value =
[| 18, 13, 16, 12 |
    20, 15, 19, 10 |
    25, 19, 18, 15 |
    16,  9, 13,  0 |] ,
```

then the following comprehension selects its second column:

```
[value[j,2] | j in 1..4] = [13,15,19,9]
```

The comprehension needed to form *vals*, a set of all the values appearing in *value* is:

```
set of int:vals = {value[j,k] | j in 1..4,k in 1..4}
```

Thus the MiniZinc data file, "assignment_data.dzn" for the 4×4 Assignment problem of Sect. 4.3.2 is as follows:

```
people = [p1,p2,p3,p4] ;
jobs = [j1,j2,j3,j4] ;
limit = 58 ;

value =
[| 18, 13, 16, 12 |
   20, 15, 19, 10 |
   25, 19, 18, 15 |
   16,  9, 12,  8 |] ;

vals = {value[j,k] | j in people,k in jobs} ;
```

8.3 Predicates and Functions

8.3.1 A Simple Predicate and a Simple Function

To keep models compact and readable, it is useful to add new constraints to those already available in MiniZinc.

The Zebra problem will illustrate the use of predicates. It is also known as "Einstein's Riddle" because it was supposed to have been invented by Einstein [57].

1. There are five houses.
2. The Englishman lives in the red house.
3. The Spaniard owns the dog.
4. Coffee is drunk in the green house.
5. The Ukrainian drinks tea.
6. The green house is immediately to the right of the ivory house.
7. The Old Gold smoker owns snails.
8. Kools are smoked in the yellow house.
9. Milk is drunk in the middle house.
10. The Norwegian lives in the first house.
11. The man who smokes Chesterfields lives in the house next to the man with the fox.
12. Kools are smoked in the house next to the house where the horse is kept.
13. The Lucky Strike smoker drinks orange juice.
14. The Japanese smokes Parliaments.
15. The Norwegian lives next to the blue house

A question posed by the riddle is "Who owns the zebra?" Try finding the answer before coding it in MiniZinc!

Assuming the houses are numbered 1..5, we need to model the constraint "next to", which is imposed in constraints 11, 12 and 15. Mathematically two houses are next to each other if their numbers differ by one. Here is a model for the problem in MiniZinc.

```
enum nationalities = {english, spanish, ukrainian, norwegian, japanese} ;
enum Colours = {red,green, ivory,yellow,blue} ;
enum animals = {dog, fox, horse, zebra, snails} ;
enum drinks = {coffee, tea, milk, oj, water} ;
enum cigarettes = {oldgold, kools, chesterfields, luckystrike, parliaments} ;
set of int: houses = 1..5 ;

array[nationalities] of var houses: Nation;
array[colours] of var houses: Colour;
array[animals] of var houses: Animal;
array[drinks] of var houses: Drink;
array[cigarettes] of var houses: Smoke;


constraint all_different(Nation) ;
constraint all_different(Colour) ;
constraint all_different(Animal) ;
constraint all_different(Drink) ;
constraint all_different(Smoke) ;
constraint Nation[english] = Colour[red] ;
constraint Nation[spanish] = Animal[dog] ;
constraint Drink[coffee] = Colour[green] ;
constraint Nation[ukrainian] = Drink[tea] ;
constraint Colour[green] = Colour[ivory] + 1 ;
constraint Smoke[oldgold] = Animal[snails] ;
constraint Smoke[kools] = Colour[yellow] ;
constraint Drink[milk] = 3 ;
constraint Nation[ukrainian] = 1 ;
constraint ((Smoke[chesterfields] = Animal[fox] + 1) \/
           (Smoke[chesterfields] + 1 = Animal[fox])) ;
constraint ((Smoke[kools] = Animal[horse] + 1) \/
           (Smoke[kools] + 1 = Animal[horse])) ;
constraint Nation[japanese] = Smoke[parliaments] ;
constraint ((Nation[norwegian] = Colour[blue] + 1) \/
           (Nation[norwegian] + 1 = Colour[blue])) ;

var nationalities: Solution ;
constraint Nation[Solution] = Animal[zebra] ;

solve satisfy;
```

The constraints involving "next to" are encoded using disjunction, for example:
```
((Nation[norwegian] = Colour[blue] + 1)
 \/ (Nation[norwegian] + 1 = Colour[blue]))
```
Instead of having to use this encoding in all three constraints, it can be declared as a *predicate*, thus:

```
predicate nextto(var houses:H1, var houses:H2) =
        (H1 + 1 = H2 \/ H1 = H2 + 1) ;
```

The two variables $H1$ and $H2$ are the arguments of the predicate, enabling it to be used in different constraints with different arguments. With this predicate it is possible to express the constraints 11, 12 and 15 more naturally:

```
constraint nextto(Smoke[chesterfields], Animal[fox])   ;
constraint nextto(Smoke[kools], Animal[horse])   ;
constraint nextto(Nation[norwegian], Colour[blue])   ;
```

Also if the constraint needs to be changed—if for example houses is changed to an *enum* instead of a set of integers, then only the predicate needs updating, rather than every constraint involving *next to*.

Just as we introduced a predicate for "next to", we can introduce a function for "immediately to the right of":

```
function var houses:rightof(var houses:H) = H+1 ;
```

and using this we can express constraint 6 as:

```
rightof(Colour[ivory]) = Colour(green) ;
```

8.3.2   Predicates and Functions with Local Variables

Predicates and functions can be much more complicated than this. On page 82 the Black-Scholes model was introduced, using two constraints on the cumulative normal distribution of the option expiry date probability. Its modelling in MiniZinc uses a function for this, illustrated in Sect. 8.5.6 below.

To illustrate what a local variable is, one can be introduced (rather unnecessarily) into the nextto predicate:

```
predicate nextto(var Houses:h1, var Houses:h2) =
        let { var int: absdiff = abs(h1-h2)} in
        absdiff= 1 ;
```

The declaration of the variable has the same syntax as declaring any other variable, except that all local definitions are wrapped in the syntax:
 let { *variable declarations* } in
Local variables can be constrained inside the *let* construct, as illustrated in this model using a function *my_av* for calculating the average:

```
function  int: my_av(array [int] of int:x) =
      let { int: y = sum(x) } in
            y div length(x) ;

var int: z ;
constraint z  = my_av([1,3,5,7,9]) ;

solve satisfy ;
```

8.3.3 Predicates for Flexibility

The constraint that all the elements in an array must be different has already appeared a few times (as in the assignment problem in Sect. 4.3.2) so it is quite familiar.

We can encode it as a predicate, expressing it as a set of disequalities. We call it *alldifferent*. The MiniZinc built-in `index_set` is used to return the set of indices of an array. This enables the same predicate to be used for any array of integers, however long it is:

```
predicate alldifferent(array [int] of var int:A) =
    (forall(i,j in index_set(A) where i<j)(A[j] != A[k])) ;
```

However it can also be modelled as a set of linear equations and inequalities. Suppose A is an array of n finite domain variables. The *alldifferent* constraint can be expressed as $\frac{(n-1)\times(n-2)}{2}$ inequalities. Using the BigM encoding each disequality $A[i] \neq A[j]$ require a boolean variable B_{ij} and two inequalities:

$$A[i] + B_{ij} \times \text{BigM} \geq A[j] + 1$$

$$A[i] + 1 \leq A[J] + (1 - B_{ij}) * \text{BigM}$$

This turns out to be very inefficient to evaluate with the linear solver described in Sect. 7.1.2.

On the other hand, we can encode A as a matrix B of 0..1 variables with $M[i, j] = 1$ whenever $A[i] = j$. The constraint that at most one of the variables $A[i]$ takes the value j is equivalent to the constraint that the sum of values in the jth column of the matrix B is 1 or 0.

The function *array2d*, which has not appeared previously, creates a matrix.
`array2d(1..2,1..3, [1,2,3,4,5,6])` creates the matrix:
```
[|1,2,3|
  4,5,6|]
```
However `array2d(1..3,1..2, [1,2,3,4,5,6])` creates the matrix:
```
[|1,2|
  3,4|
  5,6|]
```
Now we can encode the *alldifferent* constraint in the way introduced above, assuming A is an array of n variables which can only take values in the domain $min..max$. We generate the required matrix B as a function of the array A. This function is *eq_encode*:

```
function array [1..n,min..max] of var 0..1:
            eq_encode(array [1..n] of var int:A) =
        array2d(1..n,min..max, [A[i]=j | i in 1..n,
        j in min..max]) ;
```

Using the constraint on the columns of B the *alldifferent* constraint needs only m inequalities. It is encoded in MiniZinc thus:

```
predicate alldifferent(array [1..n] of var min..max:A) =
    forall(j in min..max)
          (sum(i in 1..n)(eq_encode(A)[i,j]) <= 1) ;
```

8.4 Global Constraints and Built-In Functions

The *alldifferent* constraint is one that occurs often in models.

The two ways of modelling it in the previous section are sometimes used by MiniZinc. However, as we will see later, MiniZinc can use a variety of different underlying solvers to evaluate its models. Some of these solvers have a special implementation of *alldifferent*, so rather than defining it as a predicate, MiniZinc simply passes it to the solver. Such predicates which have a special implementation in some underlying solvers are called *global constraints*. "Global" constraints are so-called because they can involve a number of decision variables, that may be different in different invocations of the constraint.

A MiniZinc-defined mapping of this constraint to its implementation in an underlying solver is available in a file called `alldifferent.mzn`, and the constraint can be included in a MiniZinc model using `include "alldifferent.mzn";` Once the mapping is included it can be used freely in a MiniZinc model. For the ecologist model of Sect. 4.3.3 we can now write:

```
int : n ;
array [1..n,1..n] of int : distances ;
int : limit ;
int : maxDist = max(i,j in 1..n)(distances[i,j]) ;
array [1..n] of var 1..n : V ;
array [1..n] of var 1..maxDist : TD ;
include "ecologist-data.dzn" ;

% all the variables in the array V  must take distinct values.
include "alldifferent.mzn" ;
constraint alldifferent(V) ;

constraint forall (j in 1..n-1) (TD[j] = distances[V[j],V[j+1]]) ;
constraint TD[n] = distances[V[n],V[1]] ;
constraint sum (j in 1..N) (TD[j]) <= limit ;
solve satisfy ;
```

The *alldifferent* global constraint automatically works out the length of the array and its indices, so it can be used on arrays of any length.

MiniZinc supports a wide variety of global constraints for different requirements, such as counting, sorting, scheduling, packing, and graphs. They are all available as predicates, for solvers which don't support them directly.

8.4.1 Two Problems Modelled with alldifferent in MiniZinc

To get a flavour of modelling with *alldifferent* in MiniZinc, here are two problems which are easy to comprehend, but hard to solve.

Sudoku

There have been many published papers on solving Sudoku's (e.g. [37, 55]). MiniZinc's *alldifferent* constraint makes it particular easy to write a Sudoku model. An *alldifferent* constraint on the *i*th row is written

```
alldifferent(j in 1..9)( puzzle[i,j] )
```
and the constraint on the *j*th column is similar. The constraint on the top-left box is written: `alldifferent(p,q in 1..3)( puzzle[p,q] )`. The constraint on the top-middle box is: `alldifferent(p,q in 1..3)` `( puzzle[3+p,q] )` and on the middle-left box it is: `alldifferent` `(p,q in 1..3)( puzzle[p,3+q] )`. An expression which imposes the *alldifferent* constraint on all nine of the boxes can thus be written:

```
forall(i,j in 0..2)
        ( alldifferent(p,q in 1..3)( puzzle[3*i+p, 3*j+q] ));
```

This example illustrates how a complex expression, like `3*i+p`, can be used to select the index of an array.

```
array[1..9,1..9] of var 1..9: puzzle;

include "alldifferent.mzn";

% All cells in a row are different.
constraint
    forall(i in 1..9)( alldifferent(j in 1..9)( puzzle[i,j] )) ;
% All cells in a column are different.
constraint
    forall(j in 1..9)( alldifferent(i in 1..9)( puzzle[i,j] )) ;
% All cells in a subsquare are different.
constraint
    forall(i,j in 0..2)
        ( alldifferent(p,q in 1..3)( puzzle[3*i+p, 3*j+q] ));

solve satisfy;
```

A Cryptarithmetic Puzzle

At the age of 13, Tom Daley became the youngest European diving champion at the Beijing Olympics. Find different digits for each letter to satisfy the following equation: "TOM × 13 = DALEY".

There are a couple of points to note when modelling this puzzle. Firstly "digits" are numbers in the range $0 \ldots 9$, but the first digit in a number can't be zero so T and D have the smaller range $1 \ldots 9$.

Secondly the value of the number *TOM* can be computed from its digits T, O and M by multiplying each digit by the required power of 10, that is $TOM = 10^2 \times T + 10 \times O + M$ note that O is the letter "O" and need not take the value zero!

Here's the MiniZinc model:

```
var 1..9 : T ;
var 0..9 : O ;
var 0..9 : M ;
var 1..9 : D ;
var 0..9 : A ;
var 0..9 : L ;
var 0..9 : E ;
var 0..9 : Y ;

include "alldifferent.mzn" ;

constraint alldifferent( [T,O,M,D,A,L,E,Y] ) ;
constraint (100*T + 10*O + M)* 13 =
            (10000*D + 1000*A + 100*L + 10*E + Y) ;

solve satisfy ;
```

The problem has just one solution, and is not easy to solve by hand.

8.5   Constraint Classes

8.5.1   Reification

Consider the disjunctive constraint $X \geq 3 \vee X = 0$. To encode this purely using integer constraints we will introduce a zero-one variable for each constraint, $B1$ and $B2$, and simply impose that $B1 + B2 \geq 1$.

Suppose that X is zero or positive, i.e. $X \geq 0$, then we can ensure that $X \leq 2 \Rightarrow B1 = 0$ through a constraint, $C1$ which in this case is: $X + (1 - B1) \times 3 \geq 3$. We can similarly ensure that $X > 0 \Rightarrow B2 = 0$, through a constraint $C2$. Now the original disjunction can be expressed as three constraints: $C1$, $C2$ and $B1 + B2 \geq 1$.

More generally, method for handling logical combinations of constraints, just using linear constraints is to formulate an integer reification of each constraint and then constrain the introduced 0, 1 variables.

Consider the finite domain constraint $Expr1 \geq Expr2$. Because this is a finite domain constraint, there is a maximum value $Max2$ that can be taken by the expression $Expr2$, and there is a minimum value $Min1$ that can be taken by the expression $Expr1$. We introduce a variable B with domain 0, 1. Let us formulate a MiniZinc constraint

```
var {0,1}:B ;
constraint Expr1+ (1-B)*(Max2-Min1) >= Expr2 ;
```

The crucial thing about this formulation is that if $Expr1 \geq Expr2$ is false, then B is forced to take the value 0.

The above toy example of a finite integer model with disjunction in MiniZinc is:

```
var 0..5: X ;
constraint X=0 \/ X>=3 ;
solve satisfy
```

It translates to the following finite integer model with conjunction only

```
var 0..5: X ;
var 0,1: B1;
var 0,1: B2 ;

constraint    % (B1=1) -> (0>=X)
  0 + (1-B1)*5 >= X ;

constraint    % (B2=1) -> (X>=3)
  X + (1-B2)*3 >= 3

% Enforce the disjunction
constraint B1+B2 >= 1 ;

solve satisfy ;
```

8.5.2 Unbounded Integer Constraints

The reason for requiring that integer expressions arc finite in this class of constraints is to ensure that models involving finite domain constraints can be solved automatically.

If unbounded expressions were allowed, then we could express Fermat's conjecture (finally proven to be a theorem 350 years later!) there is no integer N greater than 2 for which there are integers A, B and C satisfying $A^N + B^N = C^N$.

In MiniZinc this would be modelled as follows:

```
var int: N ;
var int: A ;
var int: B ;
var int: C ;
constraint N>2 /\ pow(A,N)+pow(B,N) = pow(C,N) ;
solve satisfy ;
```

Unfortunately MiniZinc just returns an error message, saying it can't iterate over an infinite set.

Naturally none of the underlying solvers for MiniZinc have the mathematical sophistication to generate Andrew Wiles' proof [56] that this model is unsatisfiable!

8.5.3 Floating Point Expressions

Suppose we are manufacturing two kinds of confectionary, butterscotch and toffee, as in Sect. 6.4.1. The first model uses floating point variables:

```
% The amount of butterscotch we make cannot be more than 5 units
% because the butter would run out.
var 0.0..5.0: Butterscotch ;
% For the same reason 5 is an upper bound on the amount of toffee
% we make
var 0.0..5.0: Toffee ;
var 0.0..40.0: Value ;

% We can only use 10 units of butter
constraint  2*Butterscotch + 2*Toffee <= 10 ;
% We can only use 10 units of sugar
constraint 1*Butterscotch + 3*Toffee <= 10 ;

constraint Value = 5*Butterscotch + 8*Toffee ;
solve maximize Value ;
```

The maximum value is in fact 32.5, achieved by making 2.5 units of butterscotch and 2.5 units of toffee. This combination uses all the available sugar, and all the butter.

By contrast if we specify the problem using a finite integer model the best solution found is 31. The finite linear model expressed in MiniZinc is as follows:

```
var 0..5: Butterscotch ;
var 0..5: Toffee ;
var 0..40: Value ;
constraint  2*Butterscotch + 2*Toffee <= 10 ;
constraint 1*Butterscotch + 3*Toffee <= 10 ;
constraint Value = 5*Butterscotch + 8*Toffee ;
solve maximize Value ;
```

8.5.4 Linear Expressions and Constraints

The class of linear constraints arise in a wide variety of problems, including production planning as exemplified above. Another nice example of modelling with linear constraints is for transportation problems. A simple transportation problem has a set of locations from which goods have to be picked up, and a set of locations where they have to be delivered as shown in Fig. 6.1 above. The complete (linear) model is as follows:

```
array [1..4] of float: dem = [2.0 , 3.8, 3.0, 1.2] ;
array [1..3] of float: cap = [5.5, 3.8, 4.2] ;
array [1..4, 1..3] of float: cost =
        [| 10.0, 7.0, 11.0 |
```

```
                 8.0,  5.0,  10.0 |
                 5.0,  5.0,   8.0 |
                 9.0,  3.0,   7.0 |] ;

array [1..4,1..3] of var 0.0..max(cap): F ;

constraint forall(i in 1..4)(sum(j in 1..3)(F[i,j])
>= dem[i]) ;
constraint forall(j in 1..3)(sum(i in 1..4)(F[i,j])
<= cap[j]) ;

var float: obj ;
constraint obj = sum(i in 1..4, j in 1..3)(F[i,j]*cost[i,j]) ;

solve minimize obj ;
```

For a linear model the optimisation expression must be linear, like all the other expressions.

8.5.5 Linear Constraints over Integers and Floats

MiniZinc can model problems involving both floating point and integer expressions. The integer expressions must have finite bounds, declared as usual. The previous transport problem becomes a MIP problem with the added constraint that a client can only be served from a single plant (introduced in Sect. 6.5).

```
array [1..4] of float: dem = [2.0 , 3.8, 3.0, 1.2] ;
array [1..3] of float: cap = [5.5, 3.8, 4.2] ;
array [1..4, 1..3] of float: cost =
          [| 10.0, 7.0, 11.0 |
             8.0, 5.0, 10.0 |
             5.0, 5.0,  8.0 |
             9.0, 3.0,  7.0 |] ;

array [1..4,1..3] of var 0.0..max(cap): F ;
array [1..4,1..3] of var 0..1: MB ;

constraint forall(i in 1..4)(sum(j in 1..3)(F[i,j])
>= dem[i]) ;
constraint forall(j in 1..3)(sum(i in 1..4)(F[i,j])
<= cap[j]) ;

constraint forall(i in 1..4,j in 1..3)(F[i,j]
<- MB[i,j]*dem[i]) ;
constraint forall(i in 1..4)(sum(j in 1..3)(MB[i,j]) <= 1) ;

var float: obj ;
constraint obj = sum(i in 1..4, j in 1..3)(F[i,j]*cost[i,j]) ;

solve minimize obj ;
```

8.5.6   *Nonlinear Problems*

Figure 6.2 above shows a simple nonlinear problem since the volume *V* of the bin
is a function of its radius *R* and its height. The height is simply the width *W* of the
mesh. The circumference of the bin $2\pi R$ is the length *L* of the mesh. The problem
can be simply modelled in MiniZinc thus:

```
int: pi = 3.14159 ;
var float:W ;                          % width of the mesh
var float:L ;                          % length of the mesh
constraint W >= 0.5 / H >= 0.5 ;       % both must be more than half
                                         a metre

var float: V ;                         % volume of the bin
var float: R ;                         % radius of the bin

constraint V = pi * R * R * W ;
constraint 2.0 * pi * R = L ;
constraint V >= 2.0 ;                  % The bin must be 2 cubic metres

solve minimize L*W ;                   % Minimize the area of the mesh
```

Using a function with local variables, we can express the Black-Scholes problem
in MiniZinc. The function `cul_normal_distr` has one arguments *D*, and four
local variables *K, A, G, Y* whose values are dependent on the input variables.
One restriction imposed by MiniZinc is the mathematical function `exp` cannot
have variables as arguments—they must be parameters. Hence RFR and MT are
declared as `float` and not as `var float`.

The function `cul_normal_distr` can be defined in two contrasting ways.
Firstly it can be defined by the following mathematical approximation:

```
function var float: cul_normal_distr(var float:D) =
    let {var float:K = 1.0 / (1.0+0.232*abs(D)),
         var float:A = K*(0.32*(-0.36+ K*(1.78+K*
         (-1.82+K*1.33)))),
         var float:G = (1.0/sqrt(2.0*pi))*exp(-D*D/2.0),
         var float: Y = abs(A*G - 0.5)} in
                  (Y + 0.5) ;
```

However this is a challenging formulation for most optimisation solving algorithms. Alternatively it can be defined by a table of values:

```
function var float: cul_normal_distr(var float: D) =
    let {var 0..409:I ,
         constraint I*0.01<=D /\ (I+1)*0.01 >= D }
    in
```

```
array1d(0..409,
[ 0.00000, 0.00399, 0.00798, 0.01197, 0.01595, 0.01994, 0.02392, 0.02790, 0.03188, 0.03586,
  0.03983, 0.04380, 0.04776, 0.05172, 0.05567, 0.05962, 0.06356, 0.06749, 0.07142, 0.07535,
  0.07926, 0.08317, 0.08706, 0.09095, 0.09483, 0.09871, 0.10257, 0.10642, 0.11026, 0.11409,
  0.11791, 0.12172, 0.12552, 0.12930, 0.13307, 0.13683, 0.14058, 0.14431, 0.14803, 0.15173,
  0.15542, 0.15910, 0.16276, 0.16640, 0.17003, 0.17364, 0.17724, 0.18082, 0.18439, 0.18793,
  0.19146, 0.19497, 0.19847, 0.20194, 0.20540, 0.20884, 0.21226, 0.21566, 0.21904, 0.22240,
  0.22575, 0.22907, 0.23237, 0.23565, 0.23891, 0.24215, 0.24537, 0.24857, 0.25175, 0.25490,
  0.25804, 0.26115, 0.26424, 0.26730, 0.27035, 0.27337, 0.27637, 0.27935, 0.28230, 0.28524,
  0.28814, 0.29103, 0.29389, 0.29673, 0.29955, 0.30234, 0.30511, 0.30785, 0.31057, 0.31327,
  0.31594, 0.31859, 0.32121, 0.32381, 0.32639, 0.32894, 0.33147, 0.33398, 0.33646, 0.33891,
  0.34134, 0.34375, 0.34614, 0.34849, 0.35083, 0.35314, 0.35543, 0.35769, 0.35993, 0.36214,
  0.36433, 0.36650, 0.36864, 0.37076, 0.37286, 0.37493, 0.37698, 0.37900, 0.38100, 0.38298,
  0.38493, 0.38686, 0.38877, 0.39065, 0.39251, 0.39435, 0.39617, 0.39796, 0.39973, 0.40147,
  0.40320, 0.40490, 0.40658, 0.40824, 0.40988, 0.41149, 0.41308, 0.41466, 0.41621, 0.41774,
  0.41924, 0.42073, 0.42220, 0.42364, 0.42507, 0.42647, 0.42785, 0.42922, 0.43056, 0.43189,
  0.43319, 0.43448, 0.43574, 0.43699, 0.43822, 0.43943, 0.44062, 0.44179, 0.44295, 0.44408,
  0.44520, 0.44630, 0.44738, 0.44845, 0.44950, 0.45053, 0.45154, 0.45254, 0.45352, 0.45449,
  0.45543, 0.45637, 0.45728, 0.45818, 0.45907, 0.45994, 0.46080, 0.46164, 0.46246, 0.46327,
  0.46407, 0.46485, 0.46562, 0.46638, 0.46712, 0.46784, 0.46856, 0.46926, 0.46995, 0.47062,
  0.47128, 0.47193, 0.47257, 0.47320, 0.47381, 0.47441, 0.47500, 0.47558, 0.47615, 0.47670,
  0.47725, 0.47778, 0.47831, 0.47882, 0.47932, 0.47982, 0.48030, 0.48077, 0.48124, 0.48169,
  0.48214, 0.48257, 0.48300, 0.48341, 0.48382, 0.48422, 0.48461, 0.48500, 0.48537, 0.48574,
  0.48610, 0.48645, 0.48679, 0.48713, 0.48745, 0.48778, 0.48809, 0.48840, 0.48870, 0.48899,
  0.48928, 0.48956, 0.48983, 0.49010, 0.49036, 0.49061, 0.49086, 0.49111, 0.49134, 0.49158,
  0.49180, 0.49202, 0.49224, 0.49245, 0.49266, 0.49286, 0.49305, 0.49324, 0.49343, 0.49361,
  0.49379, 0.49396, 0.49413, 0.49430, 0.49446, 0.49461, 0.49477, 0.49492, 0.49506, 0.49520,
  0.49534, 0.49547, 0.49560, 0.49573, 0.49585, 0.49598, 0.49609, 0.49621, 0.49632, 0.49643,
  0.49653, 0.49664, 0.49674, 0.49683, 0.49693, 0.49702, 0.49711, 0.49720, 0.49728, 0.49736,
  0.49744, 0.49752, 0.49760, 0.49767, 0.49774, 0.49781, 0.49788, 0.49795, 0.49801, 0.49807,
  0.49813, 0.49819, 0.49825, 0.49831, 0.49836, 0.49841, 0.49846, 0.49851, 0.49856, 0.49861,
  0.49865, 0.49869, 0.49874, 0.49878, 0.49882, 0.49886, 0.49889, 0.49893, 0.49896, 0.49900,
  0.49903, 0.49906, 0.49910, 0.49913, 0.49916, 0.49918, 0.49921, 0.49924, 0.49926, 0.49929,
  0.49931, 0.49934, 0.49936, 0.49938, 0.49940, 0.49942, 0.49944, 0.49946, 0.49948, 0.49950,
  0.49952, 0.49953, 0.49955, 0.49957, 0.49958, 0.49960, 0.49961, 0.49962, 0.49964, 0.49965,
  0.49966, 0.49968, 0.49969, 0.49970, 0.49971, 0.49972, 0.49973, 0.49974, 0.49975, 0.49976,
  0.49977, 0.49978, 0.49978, 0.49979, 0.49980, 0.49981, 0.49981, 0.49982, 0.49983, 0.49983,
  0.49984, 0.49985, 0.49985, 0.49986, 0.49986, 0.49987, 0.49987, 0.49988, 0.49988, 0.49989,
  0.49989, 0.49990, 0.49990, 0.49990, 0.49991, 0.49991, 0.49992, 0.49992, 0.49992, 0.49992,
  0.49993, 0.49993, 0.49993, 0.49994, 0.49994, 0.49994, 0.49994, 0.49995, 0.49995, 0.49995,
  0.49995, 0.49995, 0.49996, 0.49996, 0.49996, 0.49996, 0.49996, 0.49996, 0.49997, 0.49997,
  0.49997, 0.49997, 0.49997, 0.49997, 0.49997, 0.49997, 0.49998, 0.49998, 0.49998, 0.49998]
)[I] ;
```

Using either of these definitions the Black-Scholes equation can encoded as follows:

```
float: pi = 3.14159 ;

float: RFR = 0.1 ;
float: MT = 0.5 ;
float: R = exp(-RFR*MT) ;

var float: SP = 42.0 ;
var float: StrP = 40.0 ;
var float: SV = 0.2 ;
var float: OP ;

var float: T1 = ln(SP/StrP) ;
var float: T2 = (RFR+SV*SV/2.0)*MT ;
var float: T3 = SV*sqrt(MT) ;
var float: D1 = (T1+T2)/T3 ;
var float: D2 = D1-T3 ;
```

```
constraint OP = SP*cul_normal_distr(D1)
                -R*StrP*cul_normal_distr(D2) ;

solve satisfy ;
```

This model can itself be defined as a predicate and used in larger financial models.

8.6   Specifying the Solver in MiniZinc

The five constraint classes discussed in this chapter can be handled by a variety of different solvers available in MiniZinc [47].

The solver can be selected in the MiniZinc user interface (see Fig. 8.1) by selecting "Show configuration editor". This opens a new window pane, at the top of which is a menu of solvers visible to MiniZinc in the current installation. The default solver is normally *Gecode* [22]. Gecode handles finite domain variables, and floats, and can handle global constraints, as well as linear and non-linear constraints.

The *COIN-BC* solver [66] handles linear constraints over integers and floats. MiniZinc automatically converts other constraints into linear constraints where possible.

A third solver is *Chuffed* [14] which supports finite domain variables, but not floats. Chuffed is a "lazy clause generation" solver introduced in Sect. 9.7.2.

When MiniZinc is invoked from the command line the available solvers and their "tags" can be listed by calling `minizinc --solvers`. The chosen solver is invoked by calling `minizinc --solver <solver>`, where solver can be the name of a solver (such as `gecode`), or the name of a solver tag (such as `mip`).

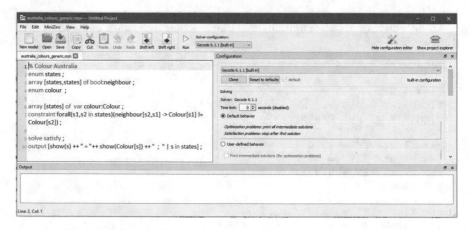

Fig. 8.1 MiniZinc solver IDE

Suppose we have a model called `model.mzn`.

If the model has finite integer expressions and constraints, it can be solved by the call

`minizinc --solver int model.mzn`.

The call for mixed integer constraints and expressions is:

`minizinc --solver mip model.mzn`

A solver that supports floating point variables can be called using:

`minizinc --solver float model.mzn`.

8.7 Summary

The MiniZinc models for the different classes of constraints use different features of the language, such as logical connectives for propositional connectives, and the floating point representation used to model real numbers. MiniZinc uses arrays and constraints on their indices to express *comprehensions*, which support compact and readable models. This chapter has introduced MiniZinc predicates and functions and local variables via the "Zebra" puzzle. Next we encountered *alldifferent* and other global constraints, as deployed for example in the "Sudoku" puzzle. With unbounded integer variables it would be possible for MiniZinc to model the Fermat conjecture—hence MiniZinc does not allow them. However MiniZinc does encode all the models of Chap. 7 including the Black-Scholes equation. Finally MiniZinc also enables users to specify which solver to deploy on their model.

8.8 Exercises

8.8.1 Trucking Exercise

There are two warehouses, A and B, each with its own truck (truckA and truckB). Each warehouse has four customers: $A1$, $A2$, $A3$, $A4$ and $B1$, $B2$, $B3$, $B4$.

There is a table of distances between each pair of customers and between the warehouses and each customer.

```
Dist =
%  A     A1      A2      A3    A4      B      B1      B2      B3    B4
[|  0,  160,    150,   590,  340,    650,   725,    560,    350,  200      % A
 |160,    0,    260,   680,  280,    650,   820,    715,    500,  150      % A1
 |150,  260,      0,   440,  260,    520,   620,    490,    280,  150      % A2
 |590,  680,    440,     0,  490,    240,   390,    435,    400,  550      % A3
 |340,  280,    260,   490,    0,    660,   800,    700,    510,  140      % A4
 |650,  650,    520,   240,  660,      0,   160,    250,    340,  670      % B
 |725,  820,    620,   390,  800,    160,     0,    210,    380,  780      % B1
 |560,  715,    490,   435,  700,    250,   210,      0,    215,  650      % B2
 |350,  500,    280,   400,  510,    340,   380,    215,      0,  450      % B3
 |200,  150,    150,   550,  140,    670,   780,    650,    450,    0 |] ; % B4
```

Assuming the trucks have a constant speed, these can also be understood as the time needed to travel between them.

First Challenge

Truck A starts at warehouse A; it serves the customers of warehouse A; and it returns to Warehouse A. Similarly for truck B.

The challenge is to minimise the time at which the latest truck returns to its warehouse.

Second Challenge

Each customer has a time window within which it must be visited, given here:

```
TimeWindow =
    [|0,  2000    %A
     |800,1800    %A1
     |100, 600    %A2
     |200,1500    %A3
     |0,   500    %A4
     |0,  2000    %B
     |200,1700    %B1
     |100,1200    %B2
     |600,2500    %B3
     |300,1700    %B4
     |] ;
```

Customer $A1$ must be visited at a time T where $800 \leq T \leq 1500$ for example. The requirement is the same as the first challenge, but additionally satisfying the time window constraints.

Third Challenge

Suppose Truck A can serve any customer and so can Truck B. The customers must be visited within their time windows. Truck A must still start and finish at warehouse A, and similarly Truck B must start and finish at warehouse B.

The challenge is to minimise the time at which the latest truck returns to its warehouse.

8.8.2   Tour Leader Exercise

A travel company has a number of tours over a season and each tour needs a tour guide. Each tour has a starting day and location, duration (days) and ending location. For the purposes of this exercise any tour guide can lead any tour, but only one at a time! Once a tour guide leads any tour he/she must be employed for the whole season (at a standard cost of 10000). When a tour guide finishes one tour and starts another there is a travel cost for going from the end location of the first tour to the start location of the next. Exactly one tour guide must be assigned to each tour.

The input is the set of tours requiring a guide, and a list of location-location travel costs. The output is a set of itineraries—one for each active tour guide—and a total cost. The parameters are specified as follows:

```
% Cost for each active tour guide
int: tour_guide_cost = 10000 ;

enum locations ;
% Each pair of locations has a travel cost recorded as an
  integer
array [locations, locations] of int: travel_cost ;

% The total number of planned tours
int: tour_ct ;
set of int: all_tours = 1..tour_ct ;
% Each tour has a start day, duration, start location and end
  location
array [all_tours] of int: tour_start;
array [all_tours] of int: tour_dur;
array [all_tours] of locations: tour_start_loc;
array [all_tours] of locations: tour_end_loc;
```

A possible data file giving the parameter values is:

```
% Example data for a toy problem instance
locations = {rome, paris, prague, munich, vienna, end} ;

travel_cost =
%       Rome, Paris, Prague, Munich, Vienna, End
    [| 0,    1106,  923,    699,    764,    0    % Rome
     | 1106,  0,    886,    685,    1034,   0    % Paris
     | 923,   866,  0,      300,    251,    0    % Prague
     | 699,   685,  300,    0,      355,    0    % Munich
     | 764,   1034, 251,    355,    0,      0    % Vienna
     | 0,     0,    0,      0,      0,      0    % End
    |]  ;

tour_ct = 7 ;
tour_start = [20,25,30,40,43,50,100] ;
tour_dur =   [15, 15, 15, 12, 10, 8, 0]  ;
tour_start_loc = [paris, paris, paris, paris, munich, munich,
end]  ;
tour_end_loc = [rome, rome, prague, munich, vienna, munich,
end]  ;
```

The tour guide cost is only incurred for guides who lead at least one tour, and each such guide leads a sequence of tours. The guide must travel from the end location of each tour in the sequence to the start location of the next, incurring the cost given in the travel cost matrix. Naturally each tour in the sequence must end before the next one starts. Every tour must have a guide.

Challenge

The challenge is to assign a guide to every tour, minimising the cost of the guides plus their travel costs.

Chapter 9
Integrating Solvers with Search

A search algorithm is one that makes choices—often by intelligent guesswork—and then explores the consequences of each choice. A complete algorithm will eventually explore all the alternative choices for each decision.

This chapter covers algorithms that use search and the different ways that the choices made during search can be exploited—either to focus on high quality candidate solutions, or to rule out choices that cannot lead to a good solution.

9.1 Generate and Test

The industrial optimisation problems addressed in this book have been those for which checking candidate solutions is relatively easy, but the huge number of candidate solutions to be checked is the challenging aspect.

Generate and test is the algorithm we can always fall back on when solving problems in this class. However there are two variants of generate and test, whose behaviour is quite different.

In the first variant complete assignments (i.e. candidate solutions) are generated and then checked to determine whether they satisfy the constraints (i.e. they are feasible). If the problem is simply to find a feasible solution, then generate and test can be lucky and the first candidate solution checked could turn out to be feasible. On the other hand generate and test can be unlucky and generate huge numbers of infeasible solutions. Indeed if there are few feasible solutions in a huge search space, finding a feasible solution by generate and test is likely to require a large computational effort.

© Springer Nature Switzerland AG 2020
M. Wallace, *Building Decision Support Systems*,
https://doi.org/10.1007/978-3-030-41732-1_9

Consider a "pigeonhole" problem, where a different hole must be assigned to each pigeon. For 50 pigeons the model needs 50 decision variables to represent the hole chosen for each pigeon. Assuming there are 50 holes, the number of candidate solutions is 50^{50}—a number(around 10^{83}) similar to the number of atoms in the universe. Generating candidate solutions randomly, the probability that a candidate is feasible is $1 * 49/50 * 48/50 * \ldots * 1/50$, which is (surprisingly) less than 0.00000000000000000001 (and can be written 10^{-20}). Consequently the number of candidate solutions that such a process can be expected to generate, before finding a feasible solution, is $1/10^{-20}$ which is a very large number (10^{20}).

There is a better variant of generate and test. Instead of generating each complete assignment as a single step, which is sometimes termed *lazy* generate and test, this version generates it via an increasing sequence of partial assignments. This can be illustrated for 3 pigeons and 3 holes, as in Fig. 9.1.

The second variant of generate and test checks each partial assignment as soon as it is created. This is termed *eager* generate and test. It removes a choice for each of the second pigeons, which has the effect of "pruning" the tree, as shown in Fig. 9.2:

Using eager generate and test for the pigeonhole problem with size 50, the worst possible choices for the nth pigeonhole would try n holes that had already been assigned, before finding an unassigned hole. Thus the eager generate and test would check $1 + 2 + \ldots + 50$ partial solutions in order to find a feasible solution. This number is merely $50*51/2 = 1275$ compared with 10^{20} for the first ("lazy") variant of generate and test!

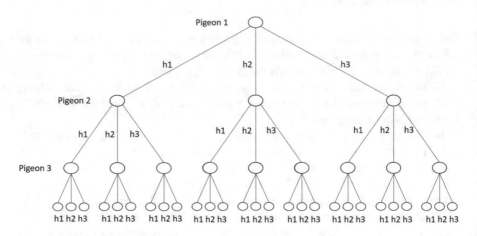

Fig. 9.1 The partial assignments tree

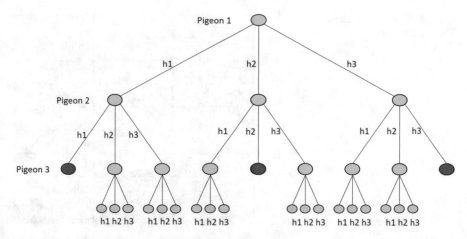

Fig. 9.2 An eager generate and test search tree

9.2   Finite Domain Propagation

The eager generate and test still checks 1275 partial solutions in the worst case. However a person would know not to try a hole that has already been assigned to a pigeon. This intelligent behaviour is captured in finite integer solvers by a mechanism known as *propagation*.

Once *pigeon1* has been assigned a hole (say $h1$) in a partial solution, propagation infers that none of the other pigeons can be assigned to $h1$. This is enforced by removing $h1$ from the domains of all the other pigeonhole decision variables.

Let's call the array of decision variables PH. The domain of $PH[i]$ is 1..50 for each index i. The pigeonhole problem imposes the constraint $PH[i] \neq PH[j]$ for every pair of distinct indices i and j.

Propagation ensures that as soon as $PH[i]$ is assigned a value h, then h is immediately removed from the domain of $PH[j]$. Consequently when a value is later assigned to $PH[j]$ it cannot be assigned any of the values previously assigned to any $PH[i]$. Thus the search for a feasible solution to the 50 pigeonhole problem requires precisely 50 steps, with every partial solution being feasible.

Propagation is nicely illustrated solving the Sudoku problem. The problem illustrated in Fig. 9.3 is a "diabolical" Sudoku problem from the age newspaper [62].

Section 8.4.1 gave a model of the Sudoku puzzle using many *alldifferent* constraints. Propagation on each of these constraints reduces the domains of certain variables. When the domain of a variable is reduced, then propagation on other constraints on the same variable seek to reduce the domains of other variables. This continues until propagation cannot further reduce any domains. Reasoning like a Sudoku player, propagation will infer that there must be a 5 in the middle of the

Fig. 9.3 Sudoku problem:
The age 26/9/2019

	5	4				9		
		6		3		2		
	9					1	7	4
	3		1		8			
8			3				4	
	7			5	9			
		7					9	5
			5		1	8		7
				9				

Fig. 9.4 Sudoku problem
after propagation

2	5	4	8	1	7	9		
7	1	6	9	3	4	2	5	8
3	9	8			5	1	7	4
	3		1		8	7		
8			3	7		5	4	
	7			5	9		8	
1		7		8			9	5
			5		1	8		7
	8		7	9			1	

top right-hand box. Knowing this, propagation, like a Sudoku player, will then infer there is another five in the bottom right corner of the middle top box, and so on.

Propagation alone on this partially complete Sudoku, yields the grid in Fig. 9.4. This is the kind of maddening state that a diabolical Sudoku leaves the puzzler in.

Propagation additionally records the set of possible values that remain for the unfilled squares in the grid, shown in Fig. 9.5. There seems no option at this point but to guess. The best approach is to guess the value in a place where there are only two possible remaining values (see below Sect. 9.4.3). The reader will be comforted to know that to solve this problem it is enough to find the value that this variable cannot take, by trying the value and performing propagation. This algorithm even has a name *singleton arc consistency*.

Fig. 9.5 Variable domains
after propagation

2	5	4	8	1	7	9	3,6	3,6
7	1	6	9	3	4	2	5	8
3	9	8	2,6	2,6	5	1	7	4
4,5, 6,9	3	2, 5,9	1	2, 4,6	8	7	2,6	2, 6,9
8	2,6	1,9	3	7	2,6	5	4	1,9
4,6	7	1,2	2,4, 6	5	9	3,6	8	1,2, 3,6
1	2, 4,6	7	2, 4,6	8	2, 3,6	3, 4,6	9	5
4, 6,9	2, 4,6	2, 3,9	5	2, 4.6	1	8	2, 3,6	7
4, 5,6	8	2, 3,5	7	9	2, 3,6	3, 4,6	1	2, 3,6

9.2.1   Arc Consistency

The idea of reducing the domains of the variables outlined in the previous section
can now be formalised. Each constraint has a set of propagators which look for
values in the domains of its variables that can be removed. A propagator can remove
a value from the domain of a variable if there are no values in the domains of
the other variables that, together with this value, satisfy the constraint. Propagation
continues until, for each variable and each constraint involving the variable:

- every value for the variable can satisfy the constraint with values in the remaining
  domain of all the other variables in the scope of the constraint

If a domain value for a variable satisfies, in this way, all the constraints in which it
occurs, we say it is *supported* by the domains of the other variables.

Definition: A set of variables and their domains is termed **arc-consistent** when
every value in the domain of every variable is supported by the domains of the other
variables [41].   □

It is interesting to note that given a set of constraints and initial domains there
is exactly one most general arc-consistent state that restricts the current domains.[1]
This is the state reached by propagation in whatever sequence the inferences on
the constraints and variable domains are performed. In short the final state is
independent of the order of propagation.

---

[1]Logically this state entails the initial domains, and is entailed by every arc consistent state that
entails the initial domains.

9.2.2   Bounds Consistency

Even though non-empty arc-consistent domains do not guarantee that there is a solution, achieving arc-consistency is still computationally expensive, especially if propagation achieving arc-consistency is performed at every node of the search tree.

Finite integer solvers often achieve a low level of consistency known as *bounds* consistency [13]. Bounds consistency is achieved by removing the highest or lowest value in the domain of each variable, until the highest and lowest values are both supported. Achieving bounds consistency therefore only requires two values in the domain of each variable to be checked for support. This avoids the number of values to be checked growing with the size of the variable domains.

Scheduling Example
Bounds consistency is often used for scheduling applications as it can achieve just the same pruning as maintaining arc-consistency on such applications [54]. Consider a toy example.

Example There are just 4 tasks, which need to be run in a week. The tasks require certain resources (resource A and/or resource B), have a release date, before which they cannot start, and a certain duration (one or two days). The resource requirements mean that two tasks using the same resource cannot run at the same time. Some tasks can only start when a previous one has finished. The task requirements are listed in Table 9.1.

This can be modelled with four decision variables $S1$, $S2$, $S3$, $S4$ whose value is the day on which the task starts.

- Given their release dates, their domains are $S1 \in 1..5$, $S2 \in 1..5$, $S3 \in 2..5$, $S2 \in 2..5$.
- Propagators check the earliest start time of each task against the precedence constraints, which updates the start times, making $S2 \in 3..5$ and $S4 \in 3..5$.
- The end times are governed by the requirement to end within the week, updating these to $S1 \in 1..4$, $S2 \in 3..4$.
- The propagator for the precedence constraint between $t1$ and $t2$ can now tighten the bound on $S1$ further: $S1 \in 1..2$.
- The propagator for the resource constraint on tasks t1 and t3 now updates the earliest start time of t3, because $S3 = 2$ is bound to clash with t1, so: $S3 \in 3..5$.
- The propagators for precedence constraint on tasks t3 and t4, tighten their end and start time respectively: $S3 \in 3..4$, $S4 \in 4..5$.

Table 9.1 Scheduling problem data

Task	Release	Duration	Resource	Preceded by
t1	1	2	A,B	
t2	1	2	A	t1
t3	2	1	A	
t4	2	1	B	t3

- Finally the resource constraint on tasks t2 and t3 rule out $S3 = 4$, so bounds propagation implies: $S1 = 1$, $S2 = 4$, $S3 = 3$, $S4 \in 4..5$.

In particular propagation infers that task t3 must run before t2 in order to complete the tasks by the end of the week.

Definition: A set of variable domains is **bounds-consistent** under a constraint when the highest and lowest values in the domain of every variable in the scope of the constraint (in other words its bounds) are supported.   □

Bounds consistent propagation achieves the widest bounds-consistent domains within the initial domains. This state is unique, so, as for arc-consistent propagation, the same final state is reached independently of the propagation order.

9.3   Tree Search

This section investigates search tree size, and different ways to search the tree. The second part of the chapter looks at a phase transition in search cost, and how optimisation impacts tree search and search cost.

Throughout the chapter we will assume that the search tree is constructed by *labelling* decision variables. A labelling search tree is one in which an uninstantiated variable is selected at each search node. There is a child node below this search node for every value in the domain of the variable. The variable is instantiated with a different value at each child node, illustrated in Fig. 9.6.

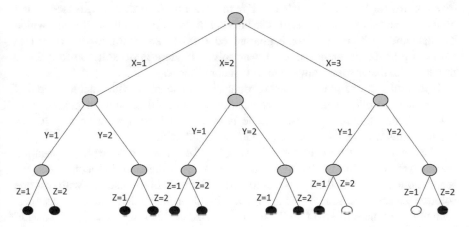

Fig. 9.6 A labelling search tree

Fig. 9.7 Search tree with
global consistency
propagation

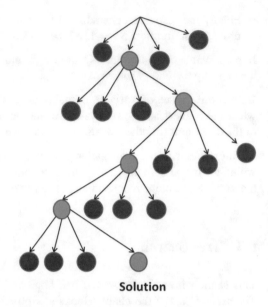

Solution

9.3.1   Complete Search Without Following a Wrong Branch

We saw in the previous chapter how constraint propagation at each search node
cuts down the size of the search tree. We noted, however, that a state (a domain for
each problem variable) could be arc-consistent even though there was no feasible
solution with the variables assigned to values in these domains. The same holds
(even more) for bounds-consistency. For this reason we term arc-consistency and
bounds-consistency *local* consistency. This contrasts with *global consistency* which
is a state under which a solution is guaranteed to exist.[2] An example of a solver that
enforces global consistency is the linear solver. If there is no solution to a linear
problem, the linear solver always reports inconsistency.

In this subsection, let us assume a propagation mechanism which achieves global
consistency. In other words, the propagation is powerful enough to guarantee to
detect whether the current search node is on a branch to a solution or not. For
example, this was the case for the pigeonhole problem above.

The required search method is quite simple. At each search node the propagation
mechanism is run and achieves global consistency. If the state is inconsistent, the
search node is failed and it has no child nodes. For a problem instance with just one
solution we show the search tree in Fig. 9.7. In this instance there are five variables,
each with a domain of size four. Note that the size of this search tree is *polynomial*
in the number of problem variables.

---

[2]Formally, there exists a solution which logically entails the current state.

By contrast, when the tree is pruned by a local consistency propagation method, in the worst case the size of the tree is exponential in the number of variables.[3]

For the first time in this chapter, let us formalise the search method for exploring this tree. We assume a global consistency propagation method which reports inconsistency (and prunes all nodes below any inconsistent node). The method is simply this.

- start at the "root" node
- if it is inconsistent then stop and report failure
- Repeat, until the current node has no children:

  - Check each child node, until finding one that is globally consistent
  - Move to that node

- Stop and return the current node (which is a solution)

Even if there are multiple solutions, this method will follow a branch towards just one solution. Assuming there are N variables and the maximum domain size is D, the number of nodes checked is at most $N \times D$.

Techniques for finding a solution after exploring as little of the original search tree as possible are crucial for solving industrial scale combinatorial optimisation problems within practical timescales.

9.3.2 Depth-First Search with Backtracking

A simple method for exploring the nodes in a search tree is termed *depth-first search with backtracking*. The method starts at the root node, moves to the leftmost child node and continues down the leftmost branch until a success or failed node is reached. On success, if only one solution is needed, the search stops. At a failure node, the method returns back up the tree to the previous node and chooses the next (second-to-the-left) child node. It then continues down the left hand branch as before.

Backtracking is required each time the method reaches a failed node. The method returns to the parent node and takes its next child. Once all the children have been explored, the method backtracks to the parent of that node, and so on until the whole search tree has been explored. The following figures illustrate a simple backtrack search. Assume all the leaf nodes are failed, so backtracking continues until the whole tree has been explored. In each diagram, the white nodes are yet to be explored; the black nodes have been completely explored; the grey nodes still have some unexplored child nodes (Fig. 9.8).

---

[3]The size of the search tree for a satisfaction problem is polynomial if there is a fixed depth *ID* such that all search nodes are guaranteed to fail at depth *ID* or less below any inconsistent node.

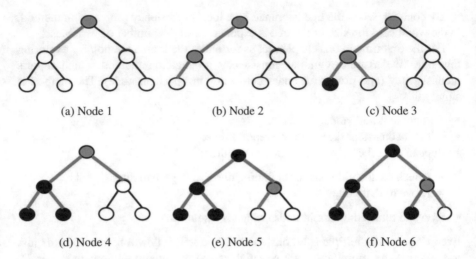

(a) Node 1   (b) Node 2   (c) Node 3

(d) Node 4   (e) Node 5   (f) Node 6

Fig. 9.8 Depth-first search with backtracking

9.4   Reordering the Search Tree

9.4.1   *Variable Order*

Propagation is one way to reduce the search tree size. Another way is simply to reorder the variables in the tree. Even with the same propagation, assigning values to the variables in a different order can change the size of the search tree. The best variable order is generally unknown, so reordering the search tree is only an intelligent guess, termed a *search heuristic*.

To illustrate this we again take an example where all the decision variables must take different values. To make the point we will make the model unsatisfiable. There are four decision variables $S1$, $S2$, $S3$, $S4$, with the following domains:

$$S1 \in 1..3,\ S2 \in 1..3,\ S3 \in 1..3,\ S4 \in 1..2$$

The search tree in which the variables are labelled in the order $S1$, $S2$, $S3$, $S4$ is shown in Fig. 9.9. Simply changing the order in which the variables are labelled, reduces the size of the search tree by a third, as shown in Fig. 9.10. In this example labelling the variable with the smallest domain first, reduced the size of the search tree. Another hint is to label first the variable involved in the most constraints—this will have a larger impact on the domains of the other variables, which ultimately reduces the size of the search tree.

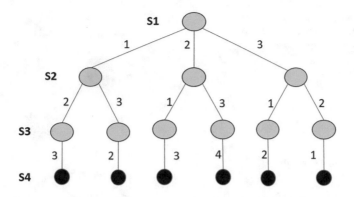

Fig. 9.9 A failed search tree

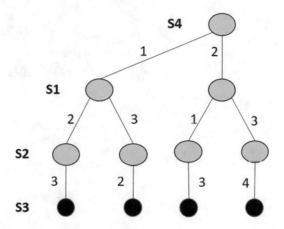

Fig. 9.10 The failed search tree re-ordered by domain size

Consider another example where the variables all have the same domain but only some of them are constrained to take different values. There are four variables $S1 \in \{red, green\}$, $S2 \in \{red, green\}$, $S3 \in \{red, green\}$, $S4 \in \{red, green\}$. There are four disequality constraints:

$$S4 \neq S1, S4 \neq S2, S4 \neq S3, S3 \neq S2$$

A nice way to view this problem is as a graph, shown in Fig. 9.11, where the nodes correspond to the problem variables, and an edge connects two nodes if there is a disequality constraint between them. A solution to the problem is a colouring of the nodes with *red* and *green* such that the two nodes joined by any edge have different colours.

Figure 9.12 shows the search tree when attempting to colour the nodes in the order $S1$, $S2$, $S3$, $S4$. The best node to colour first is $S4$ because it constrains three

Fig. 9.11 A graph to be
coloured with two colours

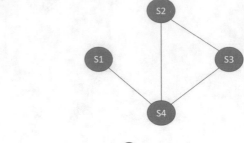

Fig. 9.12 Graph colouring
search tree

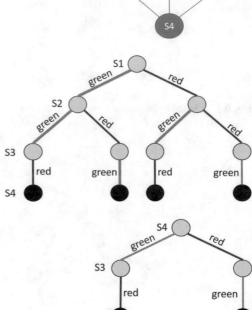

Fig. 9.13 Graph colouring
search tree reordered by
number of constraints

other nodes. Reversing the order of colouring, the search tree is much smaller, as
shown in Fig. 9.13. This is the basis of the Brelaz graph colouring order, described
and enhanced in [11]. On graphs with over 100 nodes, the choice of variable
ordering is critical: the best choice of variable can make a difference of 10 or 100
times in the number of seconds needed to colour the graph. For larger graphs the
difference is even greater.

9.4.2   Value Order

A search tree includes all the solutions of a problem. These solutions are distributed
among the leaves of the search tree. As we have seen, the exponential size of the
tree could require a number of computers greater than the number of atoms in the
universe in order to explore all the branches in parallel. Thus our assumption is that
it is not possible to explore all the branches of the search tree in parallel.

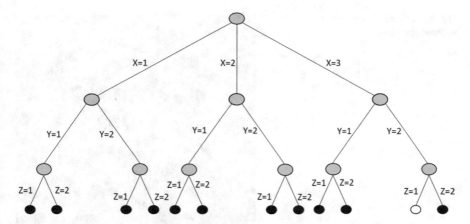

Fig. 9.14 The optimal solution discovered late in the search

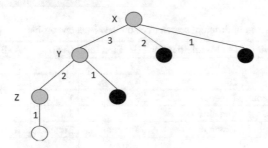

Fig. 9.15 The optimal solution discovered early in the search

Therefore the order of branches in the search tree is significant. Choosing this order is another search heuristic. Consider a toy optimisation problem with just three variables, X, Y, Z, with domains $X \in \{1, 2, 3\}$, $Y \in \{1, 2\}$, and $Z \in \{1, 2\}$ under the constraint $Y \neq Z$. The objective is maximizing $X + Y + Z$. Ordering the variables $< X, Y, Z >$ and showing the search tree with the lower values on the left yields Fig. 9.14. Assuming a branch-and-bound optimisation (described in Sect. 7.4.2), if we change the order in which values are assigned to a variable—starting with the highest value first instead of starting with the lowest, the search tree shown in Fig. 9.15.

9.4.3   Combining Search Heuristics

The impact of the shape of the search tree on the time needed to find the first solution is illustrated by the N-queens problem, where the problem is to place N queens on an $N \times N$ board in such a way that no two queens can take each other. A solution to the 8-queens problem is shown in Fig. 9.16.

Fig. 9.16 A solution to the
8-queens problem

Table 9.2 Performance of different search trees on the N-queens problem

Search steps	Naive	(A) Middle-out Var	(B) First fail	(A+B)	(A+B)+ Middle-out Val
8-queens	10	0	17	0	5
16-queens	542	28	4	0	7
32-queens	Timeout	Timeout	9	1	15
64-queens	Timeout	Timeout	365	3	2
128-queens	Timeout	Timeout	Timeout	Timeout	0

Table 9.2 give the number of search steps when finding the first solution to
different sizes of the N-queens problem under different variable and value orders.
If more than 10,000 steps are needed, the table reports *Timeout*. The model has a
variable for each column on the size N board. The value of the variable is the row
where the queen is placed.

- Naive: orders the variables from the leftmost column to the rightmost,
  $Q1, Q2, \ldots, Q8$, and orders the values from the bottom of the board to the
  top, $1, 2, \ldots 8$
- (A) Middle-out Var: orders the variables from the middle column towards the
  edges of the board: $Q4, Q5, Q3, Q6, Q2, Q7, Q1, Q8$. Values are still ordered
  from the bottom row to the top
- (B) First-fail: selects next the variable with the smallest domain. If two have the
  same smallest domain choose the leftmost one. Values are still ordered from the
  bottom row.
- (A+B): The same as first-fail, except that when two variables have the same
  smallest domain, select the one representing the column nearest the centre of
  the board.
- (A+B)+middle-out val: Same as (A+B) but order the values starting at the row in
  the middle of the board: 4, 5, 3, 6, 2, 7, 1, 8.

Labeling the queens from the middle out, and placing them, preferably, in the middle clumps the early queens together: this is the most difficult part of the search. Sooner or later queens must be placed in the middle and doing the hard part first is best— this is indeed the idea behind *first_fail*. This tackles the problem bottlenecks while there is still maximum flexibility in the remaining decisions. It avoids getting to the hard part of the problem low in the search space, such that the same cause of failure can occur again and again on different branches of the search tree. Thus it avoids, to a large extent, "thrashing" at the bottom of the search tree which causes the (dreaded!) heavy-tailed distribution of search efforts.

9.5 Methods of Exploring the Search Tree

9.5.1 Hard and Easy Problems

An easy problem is, naturally, one where it is easy (computationally cheap) to find a solution. Normally we say a problem is hard to solve if it is difficult to find a solution. However there are many problems where it is easy to recognise that there is no solution. We also term these problems easy. The hard problems, then, are ones where it is (hard to find a solution and also) hard to say if there is a solution or not.

Consider an easy problem, or model, for example one with perhaps 20 variables each with a domain size of 10, where every complete assignment is a solution. If constraints are added to the problem one at a time, then each version of the problem is a little harder than the one before because the constraints increasingly rule out candidate solutions. Eventually when many constraints have been added there may be only one or two solutions in the whole search space, whose size is 10^{20}. Add enough constraints and all candidate solutions become infeasible, and with even more constraints it becomes easy to recognise there is no solution.

Experimentally it turns out that the computation time needed to find a solution tends to increase as the number of feasible solutions decreases, peaks when there is just one or no solutions, and then reduces as more constraints are added which rule out partial assignments closer and closer to the root of the search tree, as shown in Fig. 9.17.

We term *constraint tightness* the degree to which the constraints rule out candidate solutions. The term is most suited to problems whose constraints lack any obvious structure, so that each constraint rules out a set of candidate solutions that are uncorrelated with the candidates ruled out by the other constraints.

Interestingly the graph of constraint tightness against the computation time need to find a solution has quite a sharp peak when there are very few or no solutions. The sharpness of this peak intensifies when the space of candidate solutions is large. Thus most problems, where the constraint tightness is low or high, are easy to solve. However there is a narrow band of problems with a particular constraint tightness that are computationally very costly and time-consuming to solve.

Fig. 9.17 The phase
transition between feasible
and infeasible problem
instances

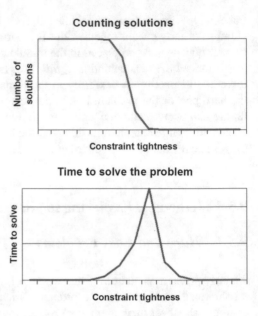

Because the number of solutions drops very sharply from a high number to none at a certain constraint tightness, and because the computation time increases dramatically at this constraint tightness and then quickly drops off again as the tightness increases, the change is known as the *phase transition* between feasible problems and infeasible problems.

To the left of the phase transition, where there are multiple solutions, the shape of the search tree can have a big impact on the computation time needed to find one. On the right of the phase transition, where there are no solutions, a good search tree enables infeasibility to be detected much faster. However at the critical constraint tightness finding the few solutions is akin to searching for a needle in a haystack. Assuming $P \neq NP$ it is arguable that no search tree shape can prevent the computation time from being impractically large for non-toy problems at the phase transition.

9.5.2 Depth-First Search

Depth-first search with backtracking was introduced in the previous section. A search method is depth-first if once a search node has been reached, the next node is one of its child nodes. The simplest form of depth-first search is *greedy* search which follows a branch until it reaches a success or failed node. Then it stops. By convention in a search tree we place the first child to be explored on the left, so a greedy search simply dives down the left branch of the tree, as in Fig. 9.18.

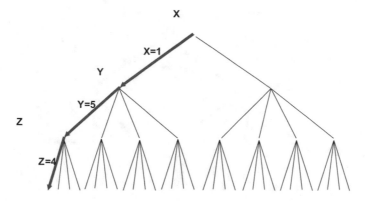

Fig. 9.18 Greedy search

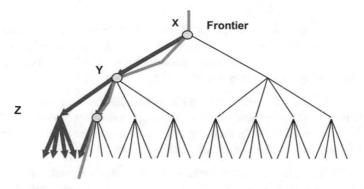

Fig. 9.19 Depth-first frontier

When alternative branches are explored on backtracking, the set of partially explored nodes (coloured grey in Fig. 9.8) is known as the *frontier*. The search method needs to keep information about all the nodes on the frontier (e.g. to record which child nodes have yet to be explored).

The advantage of depth-first search is that the partially explored nodes are always ancestors of the current node in the search tree. A consequence of this is that, in a labelling search the size of the frontier, as shown in Fig. 9.19 cannot be more than the number of decision variables.

9.5.3 Breadth-First Search

A breadth-first search method explores all the nodes at a given depth before starting to explore the nodes at the next depth. We show the *frontier* for a breadth-first search in Fig. 9.20. Breadth-first search is perfect for searching trees whose leaves are at

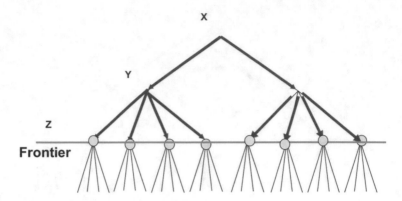

Fig. 9.20 Breadth-first search

different depths on different branches of the tree: solutions at a "shallow" depth will be found quickly. Breadth-first search also underlies a form of search called *best-first*, that focusses on the optimal solution, discussed in Sect. 9.6.2.

The disadvantages of breadth-first search are

- that in a labelling tree, where all the success nodes are at the same (maximum) depth, no solutions are explored until all the internal nodes have been expanded.
- that the number of nodes in the frontier grows exponentially with the level, and also exponentially with the number of decision variables.

The first drawback has the consequence that the time required for breadth-first search can become very large, and the second disadvantage means that the memory required (to store information about all the nodes on the frontier) grows very quickly.

9.6   Optimisation

Until now we have applied search methods to finding feasible solutions. For many—indeed most—industrial applications the requirement is not only to find a feasible solution but also a *good* one. In this case we have not only constraints, but also a measure of the cost—or quality—of each solution. For convenience we will assume in this section that the requirement is to minimise a cost. (Maximising quality can be captured as simply minimising the negative cost.)

9.6.1   *Admissible Estimates of Cost*

At a success node the cost can be evaluated precisely. However at an internal node, corresponding to a partial solution, it is also possible to assess the cost of solutions that lie on leaf nodes below it. For example if the cost is a sum of terms each of which is associated with a single variable, we can sum the terms computed from the variables which are already instantiated in the partial solution. Assuming the cost associated with a variable is always positive, the cost of any leaf node below must be greater than this partial cost of the partial solution.

We can also estimate the additional cost due to the currently unlabelled "future" variables that will be assigned values later in the search tree. An *admissible* estimate is one that is guaranteed to be less that the actual cost. In the current example, where the future variables must have a positive cost, the current partial cost $+0$ is an admissible estimate. Much of the subtlety in optimisation is in finding admissible estimates which are better than this, in that they come closer to the actual cost of the cheapest feasible solution lying under the current node.

9.6.2   Best First Search

Best-first search is a hybrid of depth-first and breadth first-search. Like depth-first search it explores the node estimated to be "closest" to the optimal solution. Like breadth-first search it maintains a frontier of partially explored nodes that may become exponentially larger than the number of decision variables.

The main idea is that when it explores a node, it examines all the child nodes recoding their cost estimates. Accordingly at any search state there is a frontier of partially explored nodes, for which the cost estimate of every child node is recorded.

At the next search step, the child node $NextNode$ with the least estimated cost is selected. The estimated cost for each of *its* child nodes is evaluated and recorded.[4] $NextNode$ is added to the frontier, and if all the children of $NextNode$'s parent have now all been selected, the parent is removed from the frontier.

If the estimated costs are admissible, then when a node with lowest estimated cost $NodeCost$ is selected, it is guaranteed that any solution to the problem must have a cost greater than $NodeCost$. Consequently when a success node (i.e. a solution) is finally selected it is guaranteed to be an optimal solution. A best-first search state is illustrated in Fig. 9.21.

Like breadth-first, best-first search can suffer from long computation times and large memory needs. Typically the admissible estimate is a gross underestimate, and node closer to the root of the tree have lower estimates than nodes closer to the leaves. Consequently most of the nodes nearer to the root of the tree are selected, and the frontier grows very large, before a success node is reached. However there are problems for which best-first search is the most efficient search method.

---

[4] According to the type of cost estimate the search is known as "Best first" or "A*".

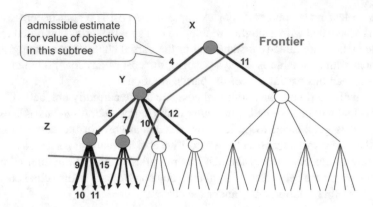

Fig. 9.21   Best-first search

9.6.3   Branch and Bound

Depth-first branch and bound is an alternative method for finding optimal solutions
quickly. It is a simple method, which starts as a depth-first search with backtracking.
When a feasible solution has finally been found, a new constraint is added to the
problem model. This is a constraint on the cost function, constraining it to take a
value better than the optimum just found.

At this point the search is either restarted from the root, with the new constraint,
or the depth-first search with backtracking continues as if the solution had been a
failed node, but again the new constraint is added to the model.

The search continues, updating the constraint on the cost upper bound, each time
a better solution is found. Eventually, if search is allowed to continue long enough,
the remaining branches of the search tree will be made infeasible by the conjunction
of the original problem constraints and the new cost constraint.

The handling of the cost constraint depends on the solver. It can be used simply
as a check when its variables are all instantiated, but it can also be used during
propagation at internal search nodes.

The advantage of depth-first branch-and-bound is that the frontier is a subset
of the ancestor nodes, as usual with depth-first search, so relatively little memory
is required. It is particularly efficient when the solver is an integer-linear solver.
The linear relaxation solved at each internal search node yields two useful kinds of
information

- A lower bound on the optimum. The lower bound is guaranteed to be lower
  than the best solution lying below this search node. If the constraint on the cost
  imposes an upper bound which is lower than this lower bound, then the node is
  failed.
- An optimal solution to the linear relaxation. The two child nodes result from
  adding constraints that exclude the current non-integer value for one of the

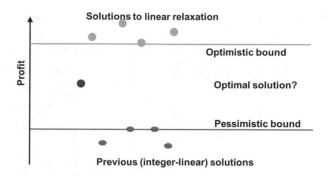

Fig. 9.22 Branch and bound

integer variables. The choice of *which* integer variable can be guided by the estimated cost of its child nodes. The linear relaxation of each of the child nodes is an admissible estimate for the cost.

Figure 9.22 illustrates optimisation with branch and bound. This gap is illustrated in Fig. 9.22.

9.6.4 Why Optimisation Is Hard

During branch-and-bound the cost bound is tightened, each time a better solution is found, until the number of feasible solutions is reduced to zero. This tightening of the lower bound, and resultant elimination of feasible solutions has precisely the effect of increasing the constraint tightness of a problem until it hits the phase transition! Thus, while problems where a feasible solution must be found can be easy or hard, *optimisation* problems are almost always hard.

The rare times that an optimisation is not hard is if there are many optimal solutions, and moreover once the cost function is constrained to be better than the optimum, the constraint tightness is past the peak and the "proof of optimality" turns out to be easy. This is illustrated in Fig. 9.23.

Note that though the optimum itself may lie to the left of the phase transition, adding the constraint that the cost is *strictly* less than the optimum can (exceptionally) drive the constraint tightness beyond the phase transition.

Fig. 9.23 Optimisation and
the phase transition

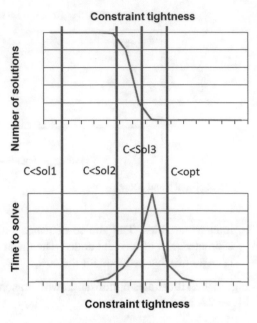

Fig. 9.24 Heavy-tailed
distribution

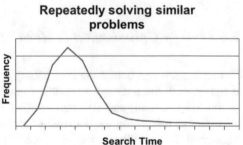

9.7   Learning and Restarts

9.7.1   Restarts

There is a surprising disadvantage to depth-first search. For "easy" problems, which
have many solutions, and whose constraint tightness is quite low, the search method
usually finds a solution quickly. However occasionally the wrong child node is
selected near the root of the search tree and the resulting subproblem turns out
to be much harder to solve than the original problem. In effect the subproblem
turns out, by chance, to lie on the phase transition. Although such cases are rare,
the computational cost of the search when it happens is extremely large. The
phenomenon of a rare event having a very large cost—so large in fact that the cost
outweighs the rarity—is termed a *heavy-tailed* distribution, illustrated in Fig. 9.24.

When sampling from a heavy-tailed distribution, the rare, extreme events are so costly that the average cost increases as more trials are carried out. The practical consequence for depth-first search methods is that even on easy problems the average time to solve a problem instance is actually dictated by the search cost at the phase transition.

A simple way to avoid the heavy-tail phenomenon is simply to restart depth-first search after a certain computation time has elapsed. Naturally the subsequent search method cannot follow exactly the behaviour of the original search, so when search restarts the variable or value selection method must be different from that used before.

A simple and practical way to achieve this is by making some choices randomly. One commonly used method is to apply a variable choice method, such as first-fail, but to break ties *randomly* when two or more variables share the same smallest domain size.

Restarting works well as a method of finding the first solution in the easy region on the left of the critical constraint tightness where the phase transition occurs. Naturally if a complete search for *all* solutions is required, the computation cost associated with the heavy-tailed phenomenon cannot be avoided: the difficult subproblem must be explored together with all the other subproblems.

For industrial combinatorial problems, therefore, search methods are typically used which, like the restart method, are not designed to explore the complete search tree.

9.7.2   *Learning*

Unit resolution alone is not enough to give the results on SAT benchmarks shown in Sect. 7.3. In addition to performing unit resolution at each node of the search tree, SAT solvers learn from their mistakes! Whenever a search node is failed, the solver derives an explanation for the failure, in the form of a minimal set of variable assignments that conflict with the last choice made during search [42].

The addition of these new clauses to the SAT problem during search has its greatest impact when search restarts. Search with restarting has a statistically better behaviour, but it also enables another advantage.

When restarting the search can focus on choices that were involved in the explanation of failures during the previous searches. This focus on the areas of conflict is termed "VSIDS" (for Variable State Independent Decaying Sum). It focusses search on the difficult choices—the problem bottlenecks, and has a huge impact in reducing the search effort required to solve most problems [48].

It was pointed out in Sect. 7.2.5 above that propositional models require a large number of constraints and decision variables, especially in order to represent numbers and numerical constraints. Finite integer models are much more compact, and intelligible. There is a solving approach called *Lazy Clause Generation* that uses restarts, learning and focussing search on the problem bottlenecks, for solving finite domain models [58].

Whenever a finite domain search reaches an infeasible partial solution, it records the reason for failure as a propositional clause. This is an example of learning while searching—the system has learnt not to try any partial solution which would be infeasible for the same reason. After a number of failures, a whole set of clauses are learnt, and VSIDS can be applied to select bottleneck variables from those appearing in the learnt clauses. The propositional variables appearing in these clauses are related to the finite domain variables in the original model by *channeling* constraints. An example of a channeling constraint relating a propositional variable B_{Vj} to the assignment of the finite domain variable V to the value j, is the (propositional) constraint $B_{Vj} = (V = j)$.

On restarting, these variables are constrained not only by the original finite domain constraints, but also by the learnt clauses.

9.8   Incomplete Search Methods

9.8.1   *Bounded Backtrack Search and Limited Discrepancy Search*

One way to avoid long search times is to limit the number of nodes explored by the search method. Once the node limit has been reached, search simply stops. In practice rather than limiting the number of nodes explored, search methods limit the number of backtracks, though the effect is just the same. The method of limiting the number of nodes, or backtracks, is termed *bounded backtrack* search.

The drawback is, of course, that if the subproblem being explored is indeed the rare, hard subproblem, then the search will fail to find a solution (even if there are a large number of solutions in other parts of the search tree).

A more subtle way of escaping from a hard subproblem is termed *limited discrepancy search*. In this search method the left-hand child node is selected first, and search proceeds down the left-hand branch of the search tree as usual. However if a solution is not found, search does not remain focussed on the "bottom-left-hand corner" of the search space as a normal depth-first search with backtracking would. Instead the search is allowed to make a non-left-hand choice at only one level in the search tree. The number of levels at which a non-left-hand branch is taken is termed the *discrepancy*. There are only $N * D$ branches with a discrepancy of 1, where D is the domain size, and so the search quite quickly tries non-left-hand branches even at the root node. The effect of this is to distribute the search across the search tree as illustrated in Fig. 9.25.

After trying a discrepancy of 1, the search method then tries a discrepancy of 2 and so on, until a solution is found.

Fig. 9.25 Limited
discrepancy search, with a
discrepancy of 1

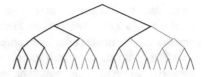

Limited discrepancy search can be neatly combined with bounded backtrack search by allowing on each branch when the allowed number of discrepancies have been incurred, a limited number of backtracks.

The number of branches with a discrepancy of K is more than D^K which, of course grows exponentially with K.[5] It is therefore often better to allow bounded backtracking and keep a smaller number of discrepancies.

Nevertheless, whatever method is chosen for exploring the search tree, the particular subtree explored can in principle capture a hard subproblem of the kind we have been seeking to avoid. That there is no guaranteed way to avoid it is a consequence of our inability to predict where solutions lie—or at least it would be if $P \neq NP$.

9.9 Summary of Tree Search

Tree search methods often work well in practice and find solutions to hard combinatorial problems. Nevertheless whichever search method is used there are drawbacks:

- Breadth-first search requires exponential time and space
- Depth-first search requires unpredictable search times
- The penalty for avoiding unpredictable search times is incompleteness

9.10 Local Search

Once all the decision variables have been assigned a value, tree search then explores other partial solutions, on the way to finding the next complete assignment. Local search, by contrast only checks complete assignments—i.e. candidate solutions. A new candidate solution is made by modifying the current assignment of some decision variables.

In case there are many constraints, it may be difficult to find an initial feasible complete assignment. Moreover it may not be possible, by modifying the current

---

[5]It is in fact $M \times D^K$, where M is the number of ways of choosing K elements from a set of size N.

candidate, to construct a new feasible candidate. Accordingly some local search methods allow constraints to be violated, but associate a penalty with each violated constraint. In this case all assignments are feasible, and satisfying the constraints comes as a result of optimisation. The *fitness* of a candidate solution comprises its objective value plus any penalties associated with violated constraints. Optimising fitness can sometimes conflict with optimising the objective. Too high a penalty on a certain constraint can drive search away from high quality solutions, but too low a penalty can leave constraints violated in the optimal solutions.

The key idea underlying local search is that the modified candidate solution is similar to the previous one, and therefore

1. the cost of the new candidate is similar to the cost of the previous one
2. the constraints satisfied in the previous assignment are mostly satisfied in the new candidate

Given one candidate solution, a local search algorithm admits a set of possible modifications to the candidate that can produce another. An example of a type of modification, is changing the value of a single decision variable. For a local search using this type of modification, on a model with N decision variables, a candidate solution can only be modified in N different ways, (by changing the value of each of the N decision variables) to produce new candidate solutions. The candidate solutions produced in this way are called the *neighbours* of the original candidate. The type of modification used by a local search is termed the *neighbourhood operator*.

The combination of the candidate solutions, their fitness values and their neighbourhoods together make up the local search *landscape*. Landscapes are often diagrammed as shown in Fig. 9.26. These diagrams are somewhat misleading, as each candidate appears to have just two neighbours—a neighbour on the left and one on the right—however they are useful for illustrating some features of local search neighbourhoods. The *global optimum* is a solution with the highest fitness value in the whole search space. In Fig. 9.26 there is just one global optimum, but in some search spaces there may be more than one.

A *local optimum* is a candidate solution none of whose neighbours are as good as the candidate. Any modification to a locally optimal candidate solution yields a candidate worse than it.

A *plateau* is a set of candidate solutions that are neighbours of each other and all have the same fitness value. Some of the solutions in a plateau will also have neighbours with higher or lower fitness values. In Fig. 9.26 this can only be true of the two candidate solutions at each end of the plateau. However in a typical problem where candidate solutions have many neighbours, many or even all, the candidates on a plateau may have other neighbours which aren't on the plateau!

There are many algorithms which employ local search, and the topic has filled many books. Useful references for this section, include the books [12] and [43] which include chapters on a variety of local search algorithms and applications.

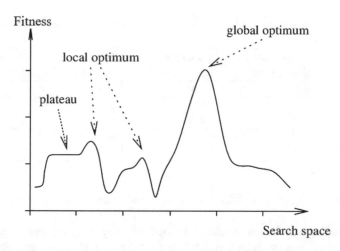

Fig. 9.26 Local search landscape

Correct Values:

	X	Y	Z
X	0	0	1
Y	0	1	1
Z	1	0	0

Candidate Solution:

X	Y	Z
0	0	0

Score: (XX:1; XY:1; XZ: 0; YY: 0; YZ: 0; ZZ: 1)   = 3

Fig. 9.27 Example problem—matching a (hidden) target

Consider the example in Fig. 9.27 which demonstrates the difference between tree search and local search.

Example In this model with N decision variables, the objective function, which is to be maximised, is the sum of a score associated with pairs of variables: the score is 1 if they each take their correct value, and 0 otherwise. The correct value (either 0 or 1) is chosen randomly for each variable in each constraint (Fig. 9.27).

Tree Search
To reach the optimal solution, with branch and bound, tree search must expand a large number of partial solutions. For 16 decision variables, branch and bound

Table 9.3 Comparing branch and bound and local improvement, averaged over 30 problems of each size

Number of variables	Av. optimum (from BandB)	Av. tree search candidate checks	Av. local optimum	Av. local search candidate checks
10	19.5	369	18.9	26
12	28.4	1,074	26.8	31
14	36.6	3,432	35.6	41
16	48.2	9,846	46.7	47
18	59.2	31,404	57.6	56

checks about 10,000 complete and partial assignments in finding the optimal solution.[6]

The branch and bound achieves limited pruning because even when some variables have been assigned a value, in a partial assignment, the total objective is still partially unknown. Since each variable is in the scope of 15 different constraints, even assigning just the last variable could improve the objective by 15.

Local Improvement

Local search starts with a complete assignment (for example setting all the variables to 0) and its neighbourhood operator just changes the value of a single variable. When any variable is changed, if the change improves the objective, the current candidate is changed and local improvement continues from this new candidate. Otherwise the algorithm seeks another neighbour of the original candidate. When the current solution has no improving neighbour (i.e. it has reached a local optimum) the search stops, and returns this candidate as the solution.

On the problem with 16 variables, the number of local search moves was around 50.[7] However the cost of the final solution returned by local improvement (the local optimum) was rarely as good as the globally optimal cost returned by branch and bound. In short there is no reason to assume that a local search has found an global optimum, though it often comes close.

A comparison of six problem instances, having an increasing number of variables and each optimised by branch and bound ("BandB") and by local improvement is shown in Table 9.3. The final solution returned from local improvement is generally close to optimal, but there is no "guarantee" of optimality. The table shows have the computational cost (measured as the number of candidate solution checks) for tree search grows steeply with the number of variables, reflecting an inherently exponential algorithm. On the other hand the computational effort for local search grows slowly, while still finding high quality solutions. This is the key difference

---

[6]This was the average number of checks in 30 runs with different randomly generated instances of the problem.

[7]Averaged over a set of 30 instances of the problem.

between *complete* algorithms which guarantee to find an optimal solution, and *heuristic* algorithms which provide no guarantees of the quality of their solutions.

9.10.1 Hill Climbing

There are many kinds of local search. The previous paragraph only accepted improving moves, a form of local search which is termed *hill climbing*. Hill climbing is the most intuitive form of local search: if a better candidate solution is found, that seems like the best place to move to. A more computationally costly form of hill climbing is to explore all the neighbours of the current candidate, and move to the best one (if it is better than the current candidate). This is termed *steepest ascent* and achieves faster improvement than just taking the first neighbour that improves on the current candidate.

Another variation of hill climbing is the behaviour on a plateau. If a hill climb accepts moves to a neighbour with the same fitness value as the current solution, then it has the chance to escape from a plateau, even if the current solution has no improving neighbours. The drawback is that the local search can, as a result, move around a plateau returning from time to time to candidates which it has encountered previously.

The limitation of hill climbing is that—as we observed in the previous paragraph—it finds a local optimum, but only rarely a global optimum. The fact that so few moves are often necessary in a hill climb like the one exampled above, is a clue to why optimality cannot be guaranteed. These 16 variable problems have 2^{16} different possible candidate solutions, but local search only checks about 50 of them!

9.10.2 Escaping from Local Optima

There are, broadly, three ways of escaping from a local optimum:

1. Moving to a neighbour with a poorer fitness
2. Generating a totally new candidate
3. Modifying the landscape itself

Moving to a Neighbour with a Poorer Fitness

A move to a worse neighbour naturally escapes the local optimum, but is as likely to be a move away from the global optimum as it is to be a move towards it. One simple way to keep moving mostly in the right direction is to accept a move to a poorer neighbour but with a lower probability than accepting a move to an improving neighbour. This way the local search should, on average, move towards improving solutions. One algorithm of this kind is *simulated annealing*.

A variant of steepest ascent is to still make a move even if the best neighbour is poorer than the current candidate solution. A consequence of this is that the next move is likely to return to the previous solution. To avoid returning to a previously encountered candidate, it is possible to record them, and reject any move to one of those previous candidates. This is the approach taken by *tabu search*. The method includes ways to avoid the list of previously encountered candidates becoming unmanageably large.

Much of the complexity of local search algorithms is in the mechanisms they use to avoid revisiting previous candidate solutions. This is termed *exploration* in contrast to standard hill climbing which is called *exploitation*.

Iterated Local Search

Iterated local search is when a local search method, such as hill climbing or simulated annealing, is allowed to run until a certain limit is reached, and then search starts again from some arbitrary new candidate solution. This is known as *restarting*. Restarting in local search is different from tree search in that the starting position is a complete assignment rather than the empty assignment in tree search. Naturally there are generic and problem-specific methods of generating a new complete assignment. Indeed tree search is such a generic method.

A form of local search which starts with a whole population of candidate solutions is termed *population-based* search. In this respect a population based search is equivalent to a local search that restarts multiple times. The extra ingredient of population-based search is the facility to combine two candidate solutions to create a new "hybrid" solution, combining parts from both parents. This facility is termed *crossover*, and is a key facility in *genetic algorithms*.

Modifying the Landscape Itself

The final way of escaping a local optimum is to change the fitness function when a local optimum is encountered. If the goal is to minimise a penalty for violated constraints, for example, then when a local optimum is encountered the penalties associated with the currently violated constraints can be increased, until the current candidate solution is no longer locally optimal. This approach is called *guided local search*.

Instead of changing the penalty, the neighbourhood can be changed. *Adaptive neighbourhood search* is an approach that can change the neighbourhoods whenever search becomes trapped in a local optimum. Neighbourhoods can be varied by simply choosing a different neighbourhood operator. For example instead of changing the value of a variable, swapping the values of two variables.

However a more drastic change is to increase the size of the neighbourhood. Instead of choosing a new value for one variable, choose new values for two variables—or indeed of three or even more variables. Naturally if a move allows all the variables to take new values, then an optimal solution is a neighbour of every candidate solution, so hill climbing with such a neighbourhood is guaranteed to yield a global optimum. Unfortunately, since the neighbourhood is the whole search space, local search simply picks candidate solutions arbitrarily and checks them, which is no better than lazy generate and test!

9.10.3   Large Neighbourhood Search

Large Neighbourhood Search (LNS) mostly avoids getting trapped in local optima just due to the size of the neighbourhood. LNS is that uses tree search to find an improving neighbour, which enables it naturally to find feasible neighbours. An LNS algorithm is specified by

- the set of decision variables whose current value is allowed to change
- the form of tree search used to assign new values to those variables

The neighbourhoods are designed to be large enough that the tree search can be constrained to only generate feasible neighbours. Thus there is no need to introduce penalties for violated constraints as required by other local search algorithms.

The advantage of LNS over tree search is that the neighbourhood search is conducted over a small proportion of the decision variables of the problem (in a large problem a "small proportion" of the decision variables may still be a large number of variables!). Consequently tree search, propagation and branch and bound can achieve good pruning of the search space (which is the set of neighbours of the current solution). Intuitively LNS selects a part of the problem and tries to resolve it better than before in the context of the rest of the solution. For its next move, LNS may select a quite different part of the solution to improve, or it might select a part of the problem which overlaps with parts which themselves have recently been improved.

Because the neighbourhoods can be large, the tree search used to find a neighbouring solution is often an incomplete tree search method such as bounded backtrack, or limited discrepancy search. For complex problems LNS may use different neighbourhood operators and different forms of tree search at each move, and may even admits moves that are not improving [52].

9.10.4   Local Search and Learning

Local search is essentially a mechanism for sampling from the search space. The mathematical basis is a technique termed *Gibbs Sampling* [23], which samples the different values for one variable against a "representative" range of assignments to all the other variables. Because most of the candidates checked have a reasonably high objective value, the sampling finds the values for the target value which are compatible with "good" assignments to the remaining variables. Since each variable in turn is a target variable when local search is exploring neighbourhoods, eventually compatible, high-value assignments are found for all the variables, even if the neighbourhood operator only changes one variable at a time.

Indeed deep learning is based on local search. The problem is modelled for deep learning by associating data values with inputs which are linked through layers of intermediate transforms in a graph to a meaningful output. Inputs flow through the graph to the output, governed by the weight attached to each edge of the graph and transforms at the nodes. Supervised deep learning is achieved by adjusting the weights of the edges on a set of training instances where the desired output is known. The weights are adjusted to weaken paths which lead to bad outputs, and strengthen paths which lead to good outputs.

The "decision variables" in deep learning are the edge weights, and the objective is the difference between the generated output and known solutions in the training set.

9.11 Summary

The chapter opens with a deeper look at the generate-and-test algorithm. The next section covers constraint propagation and the concepts of *arc* and *bounds* consistency. Some forms of search do not need to explore alternative choices, others explore alternatives during *backtracking*. Search heuristics are intelligent guesses on which choice to make next, and Sect. 9.4.3 illustrates their effect on an algorithm solving the *N-queens* problem. For a given problem class some instances have many solutions and some have none, but instances with just a few solutions are rare. These are the *hard* problems. A drawback of depth-first search is that it can lead to hard subproblems. Breadth-first and best-first search are often used in optimisation, but they also have drawbacks. These different kinds of search are enhanced by branch-and-bound, learning while searching, and restarts, all of which can retain completeness of the search. Incomplete tree search and local search are often used for large problem instances. Local search methods include hill-climbing, simulated annealing, population-based search, guided local search and large neighbourhood search, for example. Local search is used in deep learning.

Chapter 10
Search Control in MiniZinc

The default search supported by MiniZinc for its solvers is typically quite efficient. However MiniZinc also supports several ways for controlling search. In this chapter we cover variable and value ordering heuristics, and ways of combining search heuristics. The control of search is via a MiniZinc feature termed *annotation*.

10.1 The Search Annotation

The computational cost of solving combinatorial optimisation problems depends strongly on the search tree. The shape of this tree is impacted by propagation, variable ordering and value ordering. The propagation behaviour is a property of the constraint solver. MiniZinc supports a choice of solvers as outlined in Sect. 8.6.

MiniZinc enables the user to control the search by annotating the solve item in a MiniZinc model.

The syntax provided by MiniZinc is an *annotation*, which is an addition to the model that does not change the problem: logically the model remains the same whatever annotations are added to it. An annotation is written after the symbol : : An example of a search annotation is

```
solve :: int_search([X,Y,Z], input_order,indomain_min)
      satisfy
```

This annotation requests that the solver use a search method over finite integer variables: this is the meaning of int_search. The variables X, Y and Z are selected for labelling at the beginning of the search. This is the meaning of the first argument of int_search.

© Springer Nature Switzerland AG 2020
M. Wallace, *Building Decision Support Systems*,
https://doi.org/10.1007/978-3-030-41732-1_10

The second argument specifies the order in which the values should appear in the search tree. In this example input_order imposes that the three variables are selected in the order in which they are given, viz. $< X, Y, Z >$.

The final argument indomain_min specifies which choice of value for each variable will be explored next—specifically the minimum of the remaining values in the variable's domain.

10.2   Controlling Variable Order

The example of colouring a graph with two colours is modelled in MiniZinc thus:

```
enum col = {red,green} ;
array [1..4] of var col: S ;

constraint S[4] != S[1] ;
constraint S[4] != S[2] ;
constraint S[4] != S[3] ;
constraint S[3] != S[2] ;

solve
      :: int_search(S,input_order, indomain)
      satisfy ;
```

This labels the variables in the order $S[1]$, $S[2]$, $S[3]$, $S[4]$, which leads to the poor search shown in Fig. 9.14 in the previous chapter.

The best order is achieved by reversing the order of the variables in the array S, using the MiniZinc built-in function *reverse*:

```
solve
      :: int_search(reverse(S),input_order,indomain)
      satisfy ;
```

This has the same behaviour as writing:

```
solve
      :: int_search([S[4],S[3],S[2],S[1]],input_order,indomain)
      satisfy ;
```

To automatically select next the variable with smallest domain, which in this example is the best search order, use first_fail to control the variable order:

```
solve
      :: int_search(S,first_fail,indomain) satisfy ;
```

When the domain sizes are reduced during search as a result of propagation, this variable ordering takes into account the reduced domain sizes rather than the original ones. It is therefore termed a *dynamic* variable order.

For complex problems MiniZinc offers another sophisticated variable selection method called dom_w_deg. This "chooses the variable with the smallest value of domain size divided by weighted degree, which is the number of times it has been in a constraint that caused failure earlier in the search." [46].

Finally, for scheduling problems, where it is a good heuristic to assign a start time to the task which can start earliest, the variable ordering smallest is useful. This labels first the variable with the smallest value in its domain.

The scheduling example from Table 9.1 in the previous chapter can be modelled in MiniZinc as follows:

```
enum tasks = t1,t2,t3,t4 ;
enum days = mon,tue,wed,thur,fri ;

array [tasks] of days: release = [mon,mon,tue,tue] ;
array [tasks] of int: duration = [2,2,1,1] ;
set of tasks: useA = t1,t2,t3 ;
set of tasks: useB = t1,t4 ;

array [tasks] of var days:S ;

constraint forall(i in tasks)(S[i] >= release[i]) ;
constraint forall(i in tasks)(S[i]+duration[i]-1 <= fri) ;

constraint S[t2] >= S[t1]+duration[t1] ;
constraint S[t4] >= S[t3]+duration[t3] ;

include "disjunctive.mzn" ;
% Constraint on resource A
constraint disjunctive([S[i]|i in useA],[duration[i]|i in useA]) ;
% Constraint of resource B
constraint disjunctive([S[i]|i in useB],[duration[i]|i in useB]);

solve
    :: int_search(S,smallest,indomain_min)
    satisfy ;
```

The start times of the four tasks are modelled as the variable array called S. The domain of the start times is the set of weekdays. The first constraint imposes that each task must start after its release data. The second constraint imposes that each task must finish on or before Friday. Notice that the first day of the duration is the start day itself, so a task which starts on Wednesday and has a duration of two days finishes on Thursday. Adding the duration to the start day would, wrongly, give Friday.

The subsequent two constraints impose the precedences between tasks $t1$ and $t2$ and between tasks $t3$ and $t4$.

The *disjunctive* constraint is a global constraint provided by MiniZinc. It has two arguments:

- an array of start times
- an array of durations

The first disjunctive constraint enforces that tasks $t1, t2, t3$ all use resource A and therefore only one of them can be running on any day. The second disjunctive constraint enforces that tasks $t1, t4$ both use resource B, and therefore only one of them can be running on any day.

Finally the search annotation chooses to first label the start time of the only task that can run on Monday: $t1$. ($t2$ cannot start on Monday because it must be preceded by $t1$). Once the start day for $t1$ is chosen to be Monday, then $t2$ can start on Wednesday. $t3$ cannot start until Wednesday because it uses the same resource as $t1$. Thus $t4$ cannot start until Thursday because it is preceded by $t3$.

This leaves the search two alternative variables to label next: the start time of $t2$ or the start time of $t3$. This is an example of a "tie", and the solver might choose either of them to label next.

10.3   Controlling Value Order

The two most often used parameters to control the value order are `indomain_min` and `indomain_max`. The example whose search tree is illustrated in Fig. 9.14 has the following MiniZinc model:

```
var 1..3: X ;
var 1..2: Y ;
var 1..2: Z ;

constraint Y != Z ;

solve
    :: int_search([X,Y,Z],input_order,indomain_min)
    maximize X+Y+Z ;
```

Running this model in MiniZinc, solutions are printed for objective values of 4 and 5 before finding an optimal solution with an objective value of 6. Simply reversing the value order ensures the optimal is found first time:

```
solve
    :: int_search([X,Y,Z],input_order,indomain_max)
    maximize X+Y+Z ;
```

A more interesting example of value control is the N-queens problem, illustrated in Fig. 9.16. The naive n-queens model is as follows:

```
array [1..n] of var 1..n: Q;

include "alldifferent.mzn";
constraint alldifferent(Q);
constraint alldifferent(i in 1..n)(Q[i] + i);
constraint alldifferent(i in 1..n)(Q[i] - i);

solve
      :: int_search(Q, input_order, indomain_min, complete)
      satisfy ;
```

To label the variables in q from the middle out, we introduce a couple of functions:

```
% sign(k) = 1 if k is even, and sign(k) = -1 if k is odd
function int:sign(int:j) = 2*((j+1)mod 2) -1 ;

% middle_out maps 5 to [3,4,2,5,1] and maps 6 to [3,4,2,5,1,6]
  etc.
function array [int] of int: middle_out(int:n) =
    let int:mid = (n+1) div 2                in
    [mid + sign(i)*(i div 2) | i in 1..n];
```

With the function *middle_out* the middle out variable ordering is specified as follows:

```
solve
      :: int_search([Q[i]|i in middle_out(n)],input_order,
      indomain_min)
      satisfy
```

If $n = 6$ for example, this is equivalent to:

```
solve
      :: int_search([Q[3],Q[4],Q[2],Q[5],Q[1],Q[6]],input_order,
      indomain_min)
      satisfy
```

first_fail has been introduced earlier, and to combine it with middle out the search is:

```
solve
      :: int_search([Q[i]|i in middle_out(n)],first_fail,
      indomain_min)
      satisfy
```

Finally the key to being able to solve the 128-queens problem instance, we need a special variable ordering, so as to place queens earlier in the middle. The parameter for this is `indomain_median` and the search is:

```
solve
    :: int_search([Q[i]|i in middle_out(n)],first_fail,
    indomain_median)
    satisfy
```

The reason this works so well is discussed in the previous chapter. Suffice it to say that placing the early queens in the middle works for the same reason as *first_fail*.

10.4   Combining Multiple Search Variants

To achieve a different ordering for different variables, it is necessary to split the search into two parts. This is done using the MiniZinc syntax `seq_search`.

As a simple example suppose the model has two arrays of variables, X and Y, and suppose the objective is $sum(X) - sum(Y)$. Then for the X variables `indomain_max` is best, but for the Y variables we need `indomain_min`.

In general there might be constraints on the different variables, but for clarity, we can show the model with no constraints:

```
int: n ;
int: m ;

array [1..n] of var 1..n: X ;
array [1..m] of var 1..m: Y ;

solve
    :: seq_search([int_search(X,first_fail,indomain_max),
                   int_search(Y,first_fail,indomain_min)
                  ])
    maximize sum(X)-sum(Y) ;
```

Suppose, for instance, $n = 4$ and $m = 5$. Naturally the optimum solution is found immediately, but if the `indomain_min` and `indomain_max` are swapped MiniZinc prints out 32 solutions on its way to the optimum.

Essentially the second search tree is appended to the leaves of the first search tree. The first search tree, labels the X variable, with the values in left-to-right order with the maximum value in its domain first. The second search tree labels the Y variables, with the minimum values first.

A typical use of `seq_search` is to label the critical variables of a problem first, and then the less critical ones. For example in a scheduling problem it is useful to sequence the tasks on the busiest machine first.

An extended example of this is in solving a job-shop scheduling problem. In this problem there are a number of jobs to process, but each job comprises a number of tasks which have to be scheduled in the correct sequence: the next task cannot start until the previous task is finished. Finally each task must run on a specified machine—and each machine can only run one task at a time.

A classic instance of the job-shop scheduling problem is the 10 jobs, 10 tasks and 10 machines problem called "MT10" from [49] The data is as follows:

```
n_jobs = 10;
n_machines = 10;
% The rows are jobs, the columns are tasks,
% giving the machine on which this task runs
jt_machine = array2d(jobs, tasks,
        [ 0,  1,  2,  3,  4,  5,  6,  7,  8,  9,
          0,  2,  4,  9,  3,  1,  6,  5,  7,  8,
          1,  0,  3,  2,  8,  5,  7,  6,  9,  4,
          1,  2,  0,  4,  6,  8,  7,  3,  9,  5,
          2,  0,  1,  5,  3,  4,  8,  7,  9,  6,
          2,  1,  5,  3,  8,  9,  0,  6,  4,  7,
          1,  0,  3,  2,  6,  5,  9,  8,  7,  4,
          2,  0,  1,  5,  4,  6,  8,  9,  7,  3,
          0,  1,  3,  5,  2,  9,  6,  7,  4,  8,
          1,  0,  2,  6,  8,  9,  5,  3,  4,  7  ]);
% The rows are jobs, the columns are tasks,
% giving the duration of this task
jt_duration = array2d(jobs, tasks, [
          29, 78,  9, 36, 49, 11, 62, 56, 44, 21,
          43, 90, 75, 11, 69, 28, 46, 46, 72, 30,
          91, 85, 39, 74, 90, 10, 12, 89, 45, 33,
          81, 95, 71, 99,  9, 52, 85, 98, 22, 43,
          14,  6, 22, 61, 26, 69, 21, 49, 72, 53,
          84,  2, 52, 95, 48, 72, 47, 65,  6, 25,
          46, 37, 61, 13, 32, 21, 32, 89, 30, 55,
          31, 86, 46, 74, 32, 88, 19, 48, 36, 79,
          76, 69, 76, 51, 85, 11, 40, 89, 26, 74,
          85, 13, 61,  7, 64, 76, 47, 52, 90, 45
]);
```

A MiniZinc model of the problem is this:

```
% Model parameters
int: n_machines;              % The number of machines
int: n_jobs;                  % The number of jobs.
int: n_tasks = n_machines;    % Each job has one task per
                              %   machine.

set of int: jobs = 1..n_jobs;
set of int: tasks = 1..n_tasks;
```

```
set of int: machines = 1..n_machines ;

array [jobs, tasks] of machines: jt_machine;
array [jobs, tasks] of int: jt_duration;

% Time horizon - author's cheat to keep the model simple
int: max_end = 1050 ;

% Model variables.
% The start time of each job task.
array [jobs, tasks] of var 0.. max_end: jt_start;
% The finishing time is the time of the last task to complete.
var 0..max_end: t_end ;

% Constraints.
    % Each job task must complete before the next.
constraint
    forall ( j in jobs, k in 1..(n_tasks - 1) ) (
        jt_start[j, k] + jt_duration[j, k]  <=
            jt_start[j, k + 1]
    );

%       % Tasks on the same machine cannot overlap.
 include "disjunctive.mzn" ;
constraint
    forall(m in machines)
    (disjunctive([jt_start[j,t]|j in jobs,t in tasks where
                    jt_machine[j,t]=m],
                 [jt_duration[j,t]|j in jobs, t in tasks where
                    jt_machine[j,t]=m])
    ) ;

solve
    :: int_search([jt_start[j,t] | j in jobs, t in tasks],
    smallest, indomain_min)
    minimize t_end ;
```

The variable choice labels the task whose start time has the smallest domain lower bound. This model reaches a value of 1047 for the objective after 5 minutes.

A better way to label the variables is to first choose in what order the tasks are scheduled on each machine. Once this is done, every task can start as soon as its previous tasks, in the job and on the machine, are finished. Thus the start times can be labelled without any search.

To order the tasks on the machines it is necessary to introduce an array of variables denoting the sequence of tasks on each machine:

```
include "alldifferent.mzn" ;
array[machines,jobs] of var jobs: seq;

constraint
    forall(m in machines)(alldifferent([seq[m,j]|j in jobs])) ;
constraint
    forall(m in machines, j1 in jobs,j2 in jobs,t1 in tasks,
           t2 in tasks
        where jt_machine[j1,t1]=m /\ jt_machine[j2,t2]=m)
         (seq[m,j1] < seq[m,j2] ->
             jt_start[j1, t1]+jt_duration[j1, t1]
             <= jt_start[j2, t2]
         );
```

The sequential search is used to first sequence all the tasks on all the machines, and only then to label the start times:

```
solve
    :: seq_search([
        int_search([seq[m,j] | m in machines, t in tasks,
                               j in jobs
                               where jt_machine[j,t]=m ],
                    first_fail,
                    indomain_min),
        int_search([jt_start[j,t] | t in tasks, j in jobs],
                    smallest,
                    indomain_min)
                  ])
    minimize t_end;
```

Notice that the order of the jobs for each machine is in order of task number: jobs with earlier tasks running on the machine are earlier in the list.

In this case the value of the objective after five minutes is down to 980.

10.5 Restarting

To avoid the risk of search thrashing low down in the search tree, as reflected in the heavy-tailed distribution of search performance described earlier, MiniZinc offers a facility to restart search. The simplest restart is *constant*, which simply restarts at regular intervals. Because this cannot ever guarantee to search the whole tree, MiniZinc offers *linear* restarting which increases the interval between restarts adding the original constant duration onto the interval at each restart.

Finally *luby* restart [39] alternates increasing and decreasing the interval, but the longest interval does continue to grow, so also guaranteeing completeness.

The search with restart is expressed, for example, as:

```
solve
   ::   seq_search([
        int_search([ seq[m,j]  | m in machines, t in tasks,
                                 j in jobs
                                 where jt_machine[j,t]=m ],
                    first_fail,
                    indomain_random),
        int_search([jt_start[j,t]  |  t in tasks, j in jobs],
                    smallest,
                    indomain_min)
                  ])
   :: restart_constant(10000)
   minimize t_end;
```

which imposes that the search restarts after 10000 search nodes. Notice the change in the first part of the search from `indomain_min` to `indomain_random`. Naturally there's no point restarting if the new search explores exactly the same nodes as the previous search. To make sure the search is different, adding some randomness can be useful. Adding restart and randomness the objective is down to 968 within a minute.

It is possible to compare two random searches by specifying a *random seed* in the MiniZinc user interface (under "Solver Configuration"). The same random seed and search without restarting reaches 1031 after a minute.

10.6   Summary

This chapter introduced MiniZinc search annotations, to control variable and value order. The variable order can include a static order, specified in the model, and a dynamic part (using the *first-fail* parameter) whose choice depends on the behaviour of the search so far. MiniZinc also supports *sequential* search, where different parts of the search are governed by different heuristics. Finally MiniZinc also offers alternative *restart* regimes that can be chosen by the modeller.

10.7 Exercises

10.7.1 Maximize the Differences

You must put the numbers $1..n$ in order so that the sum of the differences of neighbouring numbers is maximized.

For example if n is 4, then the sum of differences of

```
[1,2,3,4]
```

is 3. However the maximum sum of differences comes from the order

```
[2,4,1,3]
```

In this case the sum of differences is 7. Note, there is another sequence of these numbers with the same sum of differences.)

Try a simple model of this against different solvers. Try and make your model solve faster.

10.7.2 Develop a Roster

Organize a roster for a tournament at a local club, or for duties at home or at work. Try and take into account people's absences, preferences or some fairness criteria. There is an example model in the Solutions section.

Chapter 11
Uncertainty

This chapter discusses two aspects of uncertainty—*probability* and *confidence*. Probability can be handled mathematically, and will be explored in detail in this chapter. Confidence is the unquantifiable relationship between the modeller and the probabilities he or she uses in reasoning under uncertainty. We can be confident and quite wrong—as well as unconfident and absolutely right. Statistics can throw up some unintuitive results which we will cover in Sect. 11.4.

11.1 What Is Uncertainty

Uncertainty is simply the lack of complete knowledge. We can forecast that an eclipse of the moon will occur on a certain date and at a certain time with pretty much complete certainty. However predicting whether we will be able to see it involves a prediction of the weather that day, which is far less certain.

Uncertainty takes two different forms. The first is can be thought of as "known unknowns". We call it *probability*, and it will be discussed in the next section. The other form is like Donald Rumsfeld's "unknown unknowns". We will term it *confidence* (or lack of it!), which is not reducible to probability. Suppose a weather forecaster predicts there will be a 80% chance of rain. If I am fully confident the forecaster is correct, then I can plan on the basis of that 80% chance. However if I lack all confidence in the forecast, the probability is of no use to me.

11.1.1 Probability

A prediction that "it will rain tomorrow" will prove to be right or wrong on the day. However, if the forecaster predicts that "there is an 80% chance of rain tomorrow". she does not actually assume or claim knowledge about the future.

© Springer Nature Switzerland AG 2020
M. Wallace, *Building Decision Support Systems*,
https://doi.org/10.1007/978-3-030-41732-1_11

The interesting question is under what conditions a prediction that "there is a 80% chance of rain tomorrow" is right or wrong. The prediction would be right if after a large number of days, before which the forecaster's knowledge was the same, on exactly 80% of those days it rained. Sadly it would be very hard to find a large number of such days—indeed the basis on which the forecast is made is different every day.

Consider now a decision made on the basis of such a prediction. Suppose the decision is whether to hold an event in the open air or indoors. If it rained, the decision to hold the event in the open air would be wrong: if it did not rain the decision would be right. However if the same decision was made on the basis of the same forecast a large number of times, and if the forecast was right, then the decision would be right 80% of the time.

The challenge in intelligent decision-making under uncertainty is not to make the right decision every time, since this could only happen by a very lucky chance. Rather the challenge is to ensure that after making many decisions the number of correct decisions is greater than the number of incorrect ones. Moreover the more knowledge the forecaster has (i.e. the closer to 100% the prediction of the chance of the forecast event happening) the greater the proportion of correct decisions that ought to be made.

Optimisation under uncertainty can be assessed in a similar way. Better decision making under uncertainty will, after a large number of decisions, yield a better total value of outcomes than poorer decision-making.

11.1.2   Confidence

The prediction of the probability of events requires a model of some kind. Our model of a dice says that when rolled it will always end up on a face; that any roll is sufficiently complex that the final position is independent of the initial state; and that the dice is completely symmetrical so no final position can be more likely than another. This is a model we all broadly accept, more or less unconsciously. Our confidence in this model is strong, because it seems reasonable, it is widely accepted, and it is rarely disputed.

In describing climate change models, Winsberg distinguishes the probabilities yielded by the models from our confidence in the models themselves [68, section 7.6]. Our confidence in the models is based on our confidence in the modellers; their access to evidence; how many other models aligned or failed to align with this one; and whether the other models were independent of it.

In the economic sphere probability is often expressed as *risk*. In this article, [25], the example is given of an admissions officer seeking to assess the probability a candidate will succeed if accepted. Ideally she would have a model that generated the probability objectively, based on historical applications and outcomes. However in creating such a model she would have to select certain characteristics of the applicants. These should be the best compromise between significance and measurability. This selection, and therefore the model would rely on her personal beliefs and choices.

In short probability can be generated objectively by a model, assuming the inputs can also be objectively measured. On the other hand confidence is subjective: our confidence in a model is based on our viewpoint. If there are objective reasons supporting our belief in a model—for example that there are multiple models which concur—then these reasons can be added into the model—for example by running a portfolio of models rather than just one. However our confidence in the extended model is ultimately subjective.

For decision making, the only form of uncertainty that can be exploited is uncertainty in the form of probabilities. By definition, uncertainty in the form of confidence—or lack of it—cannot be modelled.

11.2   Working with Probability

11.2.1   Multiplying Probability by Cost

Consider a simple yes/no decision. You are on a journey, and the decision is whether to stop for expensive fuel now, or wait for cheap fuel 60km down the road. Based on previous experience, the probability you will make it to the cheap fuel is 90%. The key question is to assign a cost for each choice:

- Buy fuel now: cost = $C1$
- Make it to the cheap fuel: cost = $C2$
- Fail to reach the cheap fuel: cost = $C3$

You would assign a high cost to $C3$, presumably!

If you choose to take the risk, the two possible outcomes occur with the following probabilities:

- With probability 0.9 you will reach the cheap fuel, and incur a cost of $C2$
- The probability of running out of fuel is $1 - 0.9 = 0.1$, and in this case it costs you $C3$

Now we can support the best decision by choosing the lesser of the following alternatives:

1. Buy now, giving a cost of $C1$
2. Do not buy now, giving a cost of $0.9 \times C2 + 0.1 \times C3$

If $C3$ is ten times more than $C1$ (which seems reasonable) then buying now is the best option, even if $C2 = 0$!

   Optimal decision-making is complicated by the fact that the decision is not
usually just yes or no. In the next example the decision could be a value rather
than yes or no—how much fuel to buy, for example. In this case the probability
of reaching the cheap fuel is a sliding scale from 90% up to 100% depending how
much fuel you buy.

   One model for this is to fix a quantity of fuel $F0$, that guarantees your car will
make it 60km to the cheap fuel. Suppose the expensive fuel costs EF per unit of
fuel, and the cheap fuel costs CF. Now buying a smaller quantity F of expensive
fuel will have the cost $EF \times F$.

   The probability P that you make it, might then be modelled linearly in terms of
the amount of expensive fuel F (less than or equal to $F0$) that you buy. If $F = 0$
then the probability is 0.9, and if $F = F0$ the probability is 1. In between these
limits we can draw a straight line, as in Fig. 11.1.

   The straight line is defined by the equation $P = 0.9 + (F/F0) \times 0.1$. We assume
that if it gets there, the car will eventually need to be filled up at least to the amount
$F0$—with some combination of expensive and cheap fuel. If it does not get there, the
additional fuel cost is absorbed in the fixed penalty cost $C3$. The optimum solution
is the minimum of:

- the cost of buying expensive fuel ($EF \times F$) plus
- the cost of (1) buying the cheap fuel ($CF \times (F0 - F)$) multiplied by the
  probability of getting there (P), plus (2) the cost of running out of fuel ($C3$)
  multiplied by its probability ($1 - P$):

$$Objective = minimum(EF \times F, \ CF \times (F0 - F) \times P + C3 \times (1 - P))$$

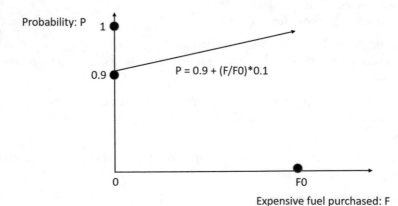

Fig. 11.1 The probability of getting to the cheap fuel

The complete model is as follows:

Parameters   – Probability of getting to the cheap fuel with no expensive fuel:
             $P0 = 0.9$
          – Amount of expensive fuel to guarantee getting there: $F0$
          – Cost of expensive fuel, per unit: EF
          – Cost of cheap fuel per unit: CF
          – Cost of running out of fuel: C3

Variables   – Quantity of expensive fuel bought: F

Constraints – The amount of fuel bought lies between 0 and $F0$: $0 \leq F \leq F0$
          – The probability of getting there is $P = P0 + (F/F0) \times (1 - P0)$

Objective   – Minimize the total cost:
             $$minimum(EF \times F, \ CF \times (F0 - F) \times P + C3 \times (1 - P))$$

In fact we are likely to overestimate $F0$ to be on the safe side. Also the estimated probability 0.9 of making it without buying expensive fuel might also be a guess. As a consequence the probabilities we read from the graph may be far from reality: in short this model might not be very good.

In this example a low confidence in the model does not mean the predicted outcomes are less likely. Indeed out model might be too pessimistic, and in reality the probability of getting to the cheap fuel might be higher than predicted by the model.

Returning to the example of climate models, having a low confidence in a climate model doesn't mean the actual probability of warming by $2°C$ is low. The model might be too optimistic, and the probability of warming is higher : - (.

11.2.2   Probability Distribution

In previous sections a probability has been associated with an event, for example the probability of it raining tomorrow, and the probability of making it to the cheap fuel.

However there is often a range of outcomes, each of which has a probability. Returning to the car running out of fuel, where the distance to the cheap fuel is 60 km. Then we might want to estimate how far short of the 60km we might run out of fuel. For this we give a *probability distribution*, as shown in Fig. 11.2. For each distance, the probability distribution shows the probability of running out of fuel at that point. The probability at distance 0 is 0. The probability is low for small distances, and grows to reach a peak at the distance the remaining fuel in the tank should normally support. Then for longer distances the probability reduces again. There is some large distance which is beyond the maximum capacity of the fuel tank, and the chance of running out of fuel beyond this distance is also 0.

We are interested in the probability the car will go further than 60km. This is given by the white area under the probability distribution. This is 0.9 as given in the problem statement. On the left of the 60 km line are the probabilities of running

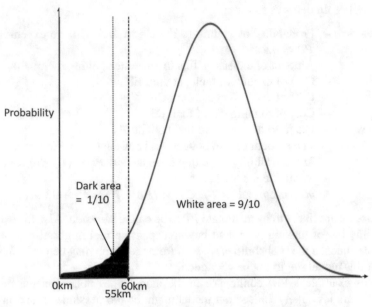

Fig. 11.2 Probability distribution for distance travelled

out of fuel after less that 60 km. The car is very unlikely to run out of fuel early in this journey, but the probability increases towards 60km and increases much more thereafter. The dark area under the curve is termed the *cumulative probability* up to 60km and a graph of the cumulative probabilities is an increasing curve termed the *cumulative probability distribution.* Its value at 60km is 0.1.

For this problem, we are not concerned with the probabilities after 60km: it is just those small probabilities on the left hand end that interest us. This part of the curve provides answers to questions like "if the car does run out of fuel, what is the chance that it is less than 5km from the cheap fuel?" This is the area under the curve between 55 and 50km, and is the difference between the cumulative probability at 60km and at 55km.

Since we do not actually know the shape of the curve the simplest model is to assume the probability decreases linearly from 60km down to 0km, as in Fig. 11.3 This is explicitly a simplification of the actual probability distribution, that we can only guess at. As before, we have to treat any solutions yielded by this model with a pinch of salt. While the model might give a precise solution, given that we have only guessed at the probability distribution, such precision belies the actual uncertainty.

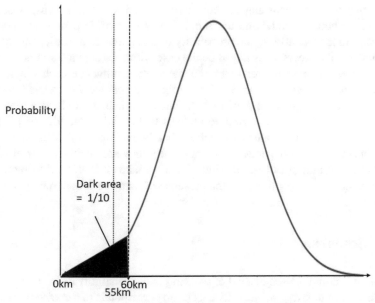

Fig. 11.3 Linear probability model for distance travelled

11.2.3   Dependent and Independent Probability Distributions

Even the very simple example above is not trivial to evaluate, and this used a very simple, linear, model of the probability of not making it to the cheap fuel.

Industrial problems include uncertainty about many factors, which creates two issues:

1. The models become more complicated and we are less confident of them
2. The payoffs become extremely complicated to evaluate

If there is uncertainty about multiple values—even if we give a probability distribution for each variable—it becomes very hard to evaluate a business decision against them all. Consider a production problem, where it has to be determined how much to produce of each of a set of products. The following parameters might be uncertain:

• user demand for each product
• raw material prices
• cost of production
• impact of advertising

A product plan might be optimal for a given probability distribution of raw material prices, given fixed demands and production costs. However when all the parameters are allowed to vary according to their own probability distributions, it becomes much harder both to model and to optimise. When multiple parameters vary they might vary independently, so the probability of one parameter is independent of the probability of the others. This might be the case with user demand and raw material prices, for example. However often the parameters influence each other, or are influenced by some common cause. They might be positively correlated, so if one parameter takes a high value then another parameter is likely to have a high value. On the other hand negative correlation means that a high value for one parameter tends to occur with a low value for another. Clearly the impact of advertising and the user demand are not independent, and hopefully they are positively correlated—at least that is the purpose of advertising. All this makes it both hard to model, and given a model, hard to optimise in the presence of many such uncertain parameters.

11.3   Scenarios

To avoid this complication instead of working with probability distributions experts propose a set of possible *scenarios*. A scenario corresponds to a choice of values for all the unknown parameters. When a set of scenarios are proposed they are assigned probabilities, such that the sum of probabilities over all the scenarios comes to 1: this reflects the assumption that exactly one of the scenarios will occur.

11.3.1   Two Scenarios

Section 11.2.1 gave an example of working with just two scenarios. This was the example of reaching the cheap fuel, or failing to get there. Each scenario had a probability:

- Scenario 1—Reaching the cheap fuel : 0.9
- Scenario 2—Failing to get there : 0.1

The choice between filling up on expensive fuel or not filling up had to be made first:

- Choice 1—Fill expensive fuel
- Choice 2—wait for the cheap fual

Finally the example had a cost for each combination. This time there are specific values for the costs: $C1 = \$100$, $C2 = \$50$, $C3 = \$350$:

Choice 1   – Scenario 1 : $100
              – Scenario 2 : $100
Choice 2   – Scenario 1 : $50
              – Scenario 2 : $350

Making the best choice requires us to take into account the probability of each scenario:

Choice 1   –   Scenario 1 : Expected cost $0.9 \times \$100 = \$90$
           –   Scenario 2 : Expected cost $0.1 \times \$100 = \$10$
           –   Combined expected cost $= \$90 + \$10 = \$100$
Choice 2   –   Scenario 1 : Expected cost $0.9 \times \$50 = \$45$
           –   Scenario 2 : Expected cost $0.1 \times \$350 = \$35$
           –   Combined expected cost $= \$45 + \$35 = \$80$

Thus, for these costs and probabilities, Choice 1 is the better choice.

11.3.2   Four Scenarios

This is a toy scheduling example, with just two tasks, $T1$ and $T2$. Each task might have two different durations, 2 hours or 4 hours. The probability of a task taking 2 hours is 40%. For 4 hours the probability is 60%. Consequently there are four possible scenarios:

Sc1 $T1$ and $T2$ each takes 2 hours: probability 0.16
Sc2 $T1$ takes 2 hours and $T2$ takes 4 hours: probability 0.24
Sc3 $T1$ takes 4 hours and $T2$ takes 2 hours: probability 0.24
Sc4 $T1$ and $T2$ each take 4 hours: probability 0.36

The choices to be made are when to run the two tasks. There are four slots available, A, B, C, D shown in Fig. 11.4, but $T1$ must run in a slot before $T2$.
Thus the possible choices are:

Ch1 $T1 \rightarrow A$ and $T2 \rightarrow C$
Ch2 $T1 \rightarrow A$ and $T2 \rightarrow D$
Ch3 $T1 \rightarrow B$ and $T2 \rightarrow C$
Ch4 $T1 \rightarrow B$ and $T2 \rightarrow D$

The costs of allocating a task to the slots is $A : \$80, B : \$50, C : \$10, D : \$100$. If a task takes too long, it may not fit the chosen slot, in which case it incurs a

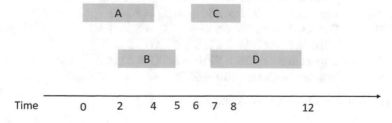

Fig. 11.4 Slots available for tasks

penalty of \$120. The lengths of the slots are $A : 4, B : 3, C : 2, D : 5$ which yields the following costs:

- T1 takes 2 hours

  $T1 \rightarrow A$ Cost: 80
  $T1 \rightarrow B$ Cost: 50

- T1 takes 4 hours

  $T1 \rightarrow A$ Cost: 80
  $T1 \rightarrow B$ Cost: 170

- T2 takes 2 hours

  $T2 \rightarrow C$ Cost: 10
  $T2 \rightarrow D$ Cost: 100

- T2 takes 4 hours

  $T2 \rightarrow C$ Cost: 130
  $T2 \rightarrow D$ Cost: 100

Finally plugging the choices into the scenarios yields the following expected costs:

Ch1   Sc1   : Expected cost $0.16 \times \$80 + \$10 = \$14.4$
         Sc2   : Expected cost $0.24 \times \$80 + \$130 = \$50.4$
         Sc3   : Expected cost $0.24 \times \$80 + \$10 = \$21.6$
         Sc4   : Expected cost $0.36 \times \$80 + \$130 = \$75.6$
         –   Combined expected cost $= \$14.4 + \$50.4 + \$21.6 + \$75.6 = \$162$
Ch2   Sc1   : Expected cost $0.16 \times \$80 + \$100 = \$28.8$
         Sc2   : Expected cost $0.24 \times \$80 + \$100 = \$43.2$
         Sc3   : Expected cost $0.24 \times \$80 + \$100 = \$43.2$
         Sc4   : Expected cost $0.36 \times \$80 + \$100 = \$64.8$
         –   Combined expected cost $= \$180$
Ch3   Sc1   : Expected cost $0.16 \times \$50 + \$10 = \$9.6$
         Sc2   : Expected cost $0.24 \times \$50 + \$130 = \$43.2$
         Sc3   : Expected cost $0.24 \times \$230 + \$10 = \$57.6$
         Sc4   : Expected cost $0.36 \times \$230 + \$130 = \$129.6$
         –   Combined expected cost $= \$240$
Ch4   Sc1   : Expected cost $0.16 \times \$50 + \$100 = \$24.0$
         Sc2   : Expected cost $0.24 \times \$50 + \$100 = \$36.0$
         Sc3   : Expected cost $0.24 \times \$230 + \$100 = \$79.2$
         Sc4   : Expected cost $0.36 \times \$230 + \$100 = \$118.8$
         –   Combined expected cost $= \$258$

The best choice is Ch1—$T1 \rightarrow A$ and $T2 \rightarrow C$.

Clearly the amount of computational effort to evaluate a set of choices against a set of scenarios increases quickly for larger problems. In the following subsections we discuss applications involving rather more scenarios.

11.3.3 Scenarios Approximating Continuous Probability Distributions

Using a finite set of future scenarios to represent uncertainty, instead of using probability distributions, is an approximation. Luckily, surprisingly few scenarios can approximate the combination of probability distributions of multiple variables. Indeed as the number of variables increases, the number of possible combinations of values increases exponentially, yet the number of scenarios to give a good approximation increases only linearly [40].

More scenarios are easier to comprehend than a combination of probability distributions, so application experts who are not necessarily experts in probability, can participate closely in the development of the scenarios. So for the production planning application some scenarios might be:

- "High growth"—high user demand, high cost of production
- "International instability"—high raw material price
- "Marketing Campaign"—high user demand
- "Business as Usual"—expected prices, costs and demands

Different scenarios are assigned different probabilities. For strategic production planning, investment in machines and equipment must be made that are robust across the likely scenarios. Later, as the scenario emerges, prices can be changed to keep demand matching production capacity as far as possible.

As illustrated above, making decisions against a set of scenarios requires each alternative decision to be evaluated against all the scenarios. Certain decisions may be ruled out because in certain scenarios they lead to unacceptable outcomes. A case where certain decisions might be ruled out occurs in the design of a product whose components have a certain reliability. The design can support more or less redundancy at a greater or smaller cost. In different scenarios different combinations of components fail, and for some design, there might be a scenario where the combined failure of certain components leads to a complete breakdown of the product. Depending on the probability associated with this scenario, this particular design must be ruled out.

Among the remaining decisions the best one is, typically, the one with the greatest expected value. The expected value of a decision is the sum of the $objective \times probability$ in each scenario.

Consider the example in Fig. 11.5, where there are 3 possible future scenarios.

In the first scenario about, the solution towards the right hand side of the graph minimises the cost, and is therefore optimal. In the third scenario, the solution

Fig. 11.5 Three alternative
future scenarios

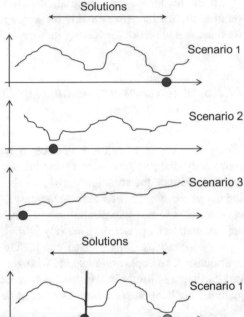

Fig. 11.6 An optimal choice
under uncertainty

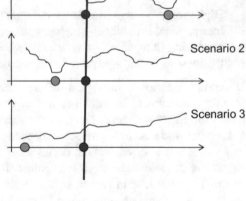

towards the left is optimal, and in the second scenario the intermediate solution
is optimal.

Naturally if the first scenario was forecast with 100% certainty, then the right-
hand decision is correct. However if the scenarios were all equally likely (each
having a probability of 1/3) then the best decision might not be any of the three
illustrated.

On the basis of its expected value,, if all three scenarios were equally likely, the
optimal choice would be that illustrated in Fig. 11.6, where the sum of the heights
of the three cost functions was minimised.

11.3.4   Solving Problems with Multiple Scenarios

The term given for solving problems with multiple scenarios is *stochastic programming*. The model for the example illustrated above requires a definition of the functions $f1$, $f2$, $f3$ and the probabilities $p1$, $p2$, $p3$ of the three scenarios (where $p1 + p2 + p3 = 1$). If X is the decision variable, the stochastic objective is simply

$$f1(X) \times p1 + f2(X) \times p2 + f3(X) \times p3$$

A more realistic example, for production scheduling, might seek the correct capacity for several machines to meet demand in future scenarios. In each future scenario the problem is to schedule the available machines so as to produce the optimal combination of amounts of several products. A model to schedule the production in each scenario can be designed: it could even be extended to generate the optimal machine capacities in that scenario. However if there were 50 or 100 scenarios, all the scenarios might use different machine capacities.

A single model combining all these models, with a single decision variable for the capacity of each of the machines in all the different scenarios, would work, but would be so large that it could not be solved. There are various methods of handling multiple scenarios without the model becoming so large as to be unsolvable in practice. They all rely on the problem belonging to certain problem classes. In fact a stochastic problem is best understood as a combination of two problems.

The first problem is to choose the values for the decision variables—these values must yield the best expected objective against all the scenarios. The second problem is to optimise the objective against a single scenario. In effect the second problem has to be solved while solving the first problem. For this reason the two problems are termed the *master* problem and the *subproblem*.

A stochastic problem is, accordingly, solved by decomposing it into a master and subproblem. How to handle this decomposition is the subject of much ongoing research [30]. A surprising result is that the set of choices which are optimal for one of the scenarios often turns out to be close to the overall optimal set of choices. This is the basis of a quite scalable approach to stochastic programming, described in [1].

A Large Scale Problem Is More Predictable
Let us return to the car rescue IDS that was introduced in Chap. 3, Sect. 3.1.1.

One technique used by one human dispatcher was to have a patrol car waiting in regions where more breakdowns tended to occur. However, because the expected number of breakdowns during the waiting period was quite small, the tactic was not worthwhile. (Though it was not easy to convince the dispatcher of this!)

In order to exploit probability effectively the number of events must be large enough for the expected number of "positive" occurrences to be a good guide to the actual number. When the number of events is small the difference between the expected number of positive events and the actual number is likely to be significant.

We illustrate this effect using a service organisation whose clients request the service on average one day in a hundred. If the organisation has a million clients then on any one day the expected number of jobs is 10,000. If each employee can carry out 10 jobs per day the organisation needs 1000 employees. If an IDSS can increase the productivity of each employee so they average 11 jobs per day, the number of employees needed is only about 900. This assumes each day has about the same number of jobs, which is indeed the case. In fact, as will be shown in Sect. 11.4.1 below, the probability that there are less than 9750 jobs in a day is less than 1%.

By contrast suppose our company has only 10,000 customers. In this case the average daily load is 100 jobs, which could be completed by 10 employees. Increasing their productivity to 11 jobs per day would suggest that an employee could be saved, as the jobs could (almost) be completed by 9 employees. However because of the smaller number of customers the total of jobs per day shows much more variance than before. Now, on roughly half the days the company should expect to have less than 94 jobs or more than 106; and one in ten days there would be less than 84 or more than 116. The consequence is that any saving from increased productivity is in effect drowned by the natural variability in the number of jobs, and little practical saving would be realised.

11.4 Statistics

We cover three statistical concepts in this section:

- sampling
- the law of large numbers
- large numbers and small probabilities

11.4.1 Sampling

Our knowledge of the real world is incomplete. This is the reason we need to deal with probabilities as discussed in the introduction to the last section. One way to learn about the real world is to reason backwards. Instead of predicting the outcome of a scenario, run the scenario; look at the outcome; and then learn what state in the real world would have made this the probable outcome.

Suppose, for example it is necessary to determine how many fish there are in a pond. Instead of emptying the whole pond and counting the fish (hopefully without killing any of them), a clever method is to take a number of fish out of the pond, mark them, and release them back into the pond again. A day or two later, when the fish have all mixed together, take a number of fish out of the pond again, and count how many are marked (Fig. 11.7). The most probable outcome is that the proportion of marked fish in the second sample is the same as the proportion of fish in the pond!

- **How many fish in the pond?**
- **Sample**
  - 30 fish
  - Mark them
  - Return them
- **Sample again**
  - 40 fish
  - There are 2 marked ones

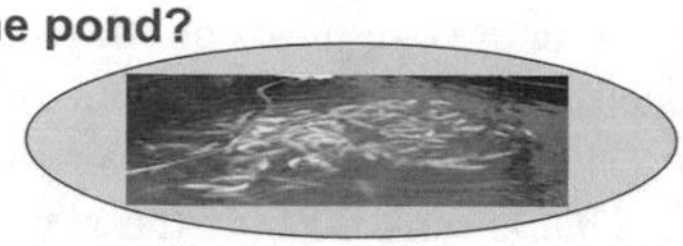

Fig. 11.7 Counting fish in a pond

Clients	Prob. of job	Average #jobs	1 day in 10	1 day in 100
1000	1 in 100	10	≤ 6 ≥ 14	≤ 3 ≥ 18
100K	1 in 100	1000	≤ 959 ≥1040	≤ 927 ≥1074
1M	1 in 100	10,000	≤ 9872 ≥10128	≤ 9768 ≥ 10232

Fig. 11.8 Engineer requests in a day

If the number of marked fish is 30, and it the second sample has a proportion of 1 marked fish in 20, then the most probable proportion of fish in the pond is 1 in 20. If there are 30 marked fish, and only 1 fish in 20 is marked, then the total number of fish must have been $30 \times 20 = 600$.

11.4.2 The Law of Large Numbers

In Chap. 3 we discussed an IDSS application for scheduling engineers to jobs at different client sites. By scheduling the jobs better it was possible to increase the number of jobs per day per employee from 7 to 8, and thus save millions of dollars per annum. The size of this benefit was due to the large number of engineers being scheduled.

However the large number has another crucial property: large numbers make things more predictable! Suppose there are three companies. Their clients are very similar: on any day each client will request an engineer's visit with a probability of 1/100. The difference between the companies is in their number of clients. The smallest company has 1000 clients, the next 100,000 and the largest has 1 million clients.

The number of engineer requests each company can expect in a day is summarised in Fig. 11.8.

The average number of requests per day increases in proportion with the number of clients: indeed it is 1/100 of the number of clients, as one would expect. The surprising thing is how often the small company has only 6 or less requests in a day: one day in 10! This is only 60% of the average number of requests. Two or three times a year there may be less than 3 requests. For such a small company, scheduling the engineers to improve their efficiency is often of no benefit since the engineers often do not have enough jobs to keep them busy.

Contrast this with the largest company. 9 days out of 10 the company has more than 9872 requests—which is over 98.7% of the average. Only one day in 100 does the number of requests drop below 97.6% of the average. The consequence is that almost all the engineers employed by the company remain busy almost all the time.

This striking predictability is a consequence of the *law of large numbers*. This law is nicely illustrated by an experiment with throwing a dice many times, as shown in Fig. 11.9. According to the law of large numbers, if a large number of dice are rolled, the average of their values is likely to be close to 3.5, with this estimate increasing in accuracy as more dice are rolled.

Suppose we run N tests (for example we ask N clients if they want a job done today). Suppose there is an expected outcome (one in ten clients say yes). We sum the actual outcomes and record the total (the number of clients who say yes). We also record the *difference between actual and expected total* call it dev_N.

The *law of large numbers* says

- given any percentage difference K (say 0.1%) and
- given any probability P (say one in ten)
- then, we can find a number N such that the probability that $dev_N > K$ is less than P

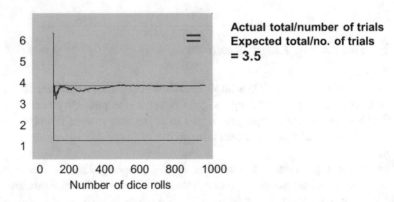

Fig. 11.9 Throwing dice—the law of large numbers

11.4.3   Large Numbers and Small Probabilities

We conclude this quick dip into statistics by considering the unintuitive outcomes which arise from large numbers and small probabilities.

We are often struck by how often coincidences occur. We see a sequence of car numbers with a specific pattern, or meet a colleague in the same cafe in a foreign town. The reason these coincidences are not signs of some mysterious force at work is that there are perhaps millions of patterns or combinations of events that would strike us as coincidences. If we were first to identify the pattern and *then* experience an event that matched it, that would be surprising. However if there were a million patterns and the chance of matching any single pattern were one in a million, then the chance of a match improves to 0.632 (not 1).

A comforting phenomena arises also from large numbers and small probabilities in medical tests. Suppose you have a test for a virus that is 98% accurate, and you test positive for the virus! This sounds worrying. Suppose, furthermore, that the virus is already widespread and 1 in 200 people have it. What are the chances you have the virus?

Suppose the chance of your having the virus and testing positive is P. Suppose on the contrary that the chance of your *not* having the virus, but still testing positive is N. Then the probability that you have the virus (given that you have already tested positive) is $P/(N + P)$.

P is the probability that you both have the virus (1/200) and tested positive (98/100), thus $P = (1/200) \times (98/100)$. N is the probability you do not have the virus (199/200) and you tested positive (2/100), thus $N = (199/200) \times (2/100)$. A quick calculation gives approximately $(P/(N + P)) = (1/5)$, so the chances that you actually have the virus are still only one in five!

11.5   Summary

The chapter started by distinguishing probability and confidence. To use probability we often multiply probability by cost in order to weight the impact of a decision. Where there are numerically distributed outcomes we need probability distributions, which are much harder to work with. Accordingly we break potential outcomes into a finite number of *scenarios*, and apply stochastic programming to reason about them. The chapter then went on, in Sect. 11.4, to discuss statistics. Statistics enable us to calculate probabilities, which we can do by sampling in case we cannot access all the data. The law of large numbers gives us confidence as our data size grows. However, Sect. 11.4.3 showed that when we have large numbers, we can easily be fooled by small probabilities.

Chapter 12
IDSS for Optimisation: The Future

Intelligent Decision Support Systems for Optimisation are growing in importance as the issues of problem specification, scalability of solvers, and flexibility of solutions are overcome. In particular the integration of machine learning, forecasting and dynamic optimisation will become important in all kinds of decision-making.

This chapter will explore some of these application areas and technologies.

12.1 Transport and Congestion

Cities all over the world are suffering from higher congestion, increasing travel times and frequent major disruptions. Travelling from Jakarta to the airport can take 3 hours, and Sao Paulo's regular commute times for its poorer citizens is 4 hours [29]. It is now recognised that building more roads simply enables people to commute from further away, increasing the traffic until the roads are as congested as before [45].

Rather than building more cars (driven and driverless) and more roads, the only practical way to address congestion is to maximise our use of the available streets. This requires each journey to be made as efficiently as possible, at the same time ensuring that all the journeys across a city are coordinated. Each journey may involve a combination of one or more modes—walking, car, bus, tram, train, and even ferry—with the minimum of waiting and travelling time, and departing or arriving to the specified deadline.

Routing apps already enable drivers to avoid congested roads, and public transport apps enable travellers to find good connections. However today's routing apps can only reflect current congestion, not how things will be in 15 minutes, and they suggest the same diversions so the congestion simply moves to another place.

Today's public transport apps do not enable transport operators to configure their fleets to meet current demand. These limitations will be overcome in the optimised systems of the future.

Optimisation will ensure the bus meets the train and there is space on both to meet demand. Optimisation will ensure that the flow on any street at any time of day is maximised. Based on incentives to travel in high-capacity vehicles at a time when the capacity is available, optimisation will break down congestion leading to faster travel times for all. In Sao Paulo, for example, 82% of drivers said they would switch to public transport if the service was better [29].

12.2 Energy

In a renewable energy future, they say the marginal cost of energy will be zero—though the infrastructure will have to be paid for [38]. However getting there will require the development of new kinds of energy networks. Current networks deliver power from a small number of high powered generators over a grid comprising major *transmission* lines, linked to smaller *distribution* lines and finally to individual energy users. The future network will link a large number of renewable energy generators, including *prosumers* who both consume energy and produce it from their own small-scale renewable sources such as solar panels.

Currently energy consumption is unpredictable, and the energy generated must be tuned to match demand. In the zero marginal energy cost future, both supply and demand will become unpredictable, and until there is long-term battery storage for large populations, demand management will become an important tool in match supply and demand.

Distributed optimisation techniques will scale to coordinate demand across cities with millions of energy users. Energy grids will be finely controlled to resiliently handle continual variations in supply.

12.3 Automation

Driverless cars have been a topic of intense discussion in the media, but meanwhile automation is proceeding apace in production, mining, the military, banking, and increasingly across the service industries. Areas currently on the brink of automation include health—with automated and remote monitoring and diagnosis—and teaching where online teaching is supported by automated tailored interaction with students.

Surprisingly automation does not on its own improve efficiency. The motivation is often safety, consistency and reliability. However a hidden benefit of automation is its responsiveness to control.

In a real-life example from a car breakdown company, the dispatching process had two objectives: minimising the response time—the time until a breakdown vehicle arrived on the site—and minimising costs. Contractors could be called on when the company's patrols were all busy, but they incurred a cost. On days when costs were too high, the manager would request the dispatchers to reduce costs, but this resulted in a massive blow-out in response times. It was not possible to say exactly when contractors should be called, and when to have a customer wait longer. A thermostat keeps temperature nearly constant by reacting whenever the temperature becomes too low or too high, but people cannot be repeatedly told to modify their behaviour like this.

Car drivers are unpredictable: sometimes they accelerate fast when the light goes green, and sometimes they pause and then pull away slowly. Sometimes they hover looking for a parking place, and sometimes they recklessly overtake. One outcome is the "phantom traffic jam" which occurs when a minor event like a truck changing lanes causes cars to brake resulting in a wave of braking that propagates back miles up the motorway. Automated vehicles have fast responsiveness that not only averts accidents but enables the flow of traffic to be resistant to such phantom jams, not only on the motorway but in urban streets as well [24].

Automated vehicles will enable a far greater number of vehicles to flow freely down a street without the usual slowdown caused by congestion. Nevertheless, as pointed out in Sect. 12.1, an automated bus can carry the same number of passengers as 50 single-occupancy cars, taking up a fraction of the road space!

The predictability of automated systems not only enables more fine-grained control, but it also enables larger systems to be integrated. Consider the unloading of a ship. Containers must be lifted off the ship by a crane (in a controlled order to avoid the ship bending under stress, or even tipping over). The unloaded containers must then be moved by straddle cranes to an area from where they will be collected and placed on a train or truck. The trucks and trains must also be scheduled to arrive at the right time and place to fetch their container. Due to the unpredictability of the time to unload and move these actions cannot be coordinated. Indeed ports the world over have mountains of containers awaiting collection, and queues of trucks waiting for their loads.

Automation enables a plan to be programmed which will be followed to the letter by the automated cranes and straddles, such that every container will have a predictable location at any time. This fine-grained control enables highly optimised plans to be computed that has the right truck, train, straddle, or crane to be in the right place exactly at the time needed for the next stage in the process.

12.4 Integrated Supply Chains

Automation of customer ordering through integrated IT systems can enable the optimiser to have a complete view of the supply chain from raw material supply and inventory, through production, and transport right to the customer location.

Production can be tailored precisely to meet demand, and raw material supply matched precisely to production, with transport scheduled so as to deliver everything on time while minimising the number of vehicle kilometres travelled.

The whole supply chain can be optimised and executed without any human intervention. The sharing of data across an organisation and between organisations enables fully integrated planning. Indeed the only limitation is the computational effort needed to handle the size of the integrated optimisation problem!

There have been years of research and development to tackle the optimisation of logistics for one organisation. Advanced optimisation can typically cut 10% of cost and carbon emissions from even an efficiently managed fleet. However the logistics of multiple organisations, with automated vehicles and IoT (the internet of things), can be integrated—equitably scheduled and fully monitored. The benefits of optimising the combined logistics of multiple organisation is closer to 25%.

In a world where almost all buying and selling—business to business, or person to person, wholesale or retail—is done online, optimised logistics will be at the heart of tomorrow's economy.

12.5   4th Industrial Revolution

An integrated automated economy will be the outcome of the 4th industrial revolution. Repetitive work will no longer be needed. As long as things run smoothly, when you or I request a product it will automatically be produced, delivered and invoiced without any human interaction.

Naturally things will not run smoothly! The role of the manager in this new world will not be to oversee a production process or a service. Instead her role will be to respond to exceptions. This raises a major challenge in the design and running of the new automated processes. The human must be able to look into the running of the system, understand what is happening and why, and to interact with it so as to successfully handle exceptions.

The integration of human factors in such complex systems is essential and cannot be added as an afterthought. An optimised solution must be explainable. It must be possible to make specific changes to this solution with minimal knock-on effects. The person interacting with the system must be the person with the appropriate expertise and permissions. The security of such complex systems is a critical issue. Today it is social media that is manipulated by crooks and evil regimes. Viruses have been planted to bring down power plants and power networks. Bringing a fully integrated 4th generation economy to a standstill would be an even greater disaster.

In short automation needs optimisation, and optimisation needs impregnable security and sophisticated support for the human in the loop. These are no small demands.

12.6 Learning, Optimisation, and Policies

The technical challenges for intelligent decision support include scalability, explainability, uncertainty, and adaptability. Any advance in one area—for example the design of a more sophisticated algorithm—raises new challenges for the other areas—how can this be explained? how can it deal with uncertainty? how can it be adapted for new situations?

There has been a paradigm shift in machine learning over the last decade. By playing Go against itself a machine racked up 4.9 million games in three days. Just remembering 4.9 million mistakes, let alone learning from them would be an achievement. Yet the number of possible moves in Go is enormous. After the first two moves of a Chess game, there are 400 possible next moves. In Go, there are close to 130,000. After a dozen it exceeds the number of atoms in the universe. What is learnt is not just to avoid making the same mistake again, but to recognize every set of patterns that arises in a new game and the right move to make in the presence of that particular set. This pattern-based choice we term a *policy*.

Most decisions that organisations make are based on incomplete information: how much raw material to buy; how many staff to employ; where to set up a new service centre. For these decisions we do not know enough about the future to guarantee their consequences. Some decisions—like daily staffing needs at a supermarket—may occur often enough (though not, perhaps, 4.9 million times!) that learning from the past can support a good policy. However a policy depends on the future being like the past, and the whole challenge of running a team, a department, a site or a company, is that the future will not be a repeat of the past.

Intelligent Decision Support for optimisation must be able to take advantage of learning and forecasting, where there is sufficient evidence to recognise patterns and trends. However it must also be able to reason about the consequences of possible changes. Supposing the cost of a raw material doubled; supposing a customer market was closed through a new regulations; supposing the impact of a marketing campaign started a huge market opportunity in a new country—what would be the best strategy?

The interplay between repetition and change, between trends and random events, between policies and one-off decisions; and between strategy, tactics and operational control—these can only be faced with a combination of human and artificial intelligence.

12.7 Summary

This chapter has discussed some big societal problems that could be addressed by intelligent decision support. Congestion in today's cities is one target for advanced and scalable decision-support. Renewable energy is vital to the future of the world, but brings with it unpredictability of energy supply. This can be solved with an infrastructure that uses advanced and comprehensive decision support. Automation is a threat to jobs, but also an opportunity to do things better—for example using our limited road space more efficiently. Indeed with automation and sophisticated decision support it would be possible for customers to place orders online and have the whole supply chain, from raw materials to production to transport and delivery, work without delay or human intervention, except after a disruption. This takes us into the 4th industrial revolution which could offer people freedom and time to follow their own interests—as long as the benefits of automation are not captured by the few at the expense of the many.

Appendix A

A.1 Big O Notation

When writing down the complexity of an algorithm, we name the parameter of interest, and how the computing time grows with the size of this parameter. We use a special notation called "Big O" notation because it uses a big "O" in the notation ☺ We write $O(X)$ to denote the complexity class identified by X.

Here is a complete list of the complexity classes which we will encounter in this text, and the associated big O notation: The term "linearithmetic" is not widely used (though it appears in Wikipedia), but the associated complexity class $O(N \log N)$ is one that occurs quite often (Table A.1). In particular the best sorting algorithm has this complexity (where N is the length of the list to be sorted).

The Big O notation denotes an approximation. For example the computing time for a linear algorithm need not increase in exact proportion to the problem size N but it could be in proportion to $C \times N$ for any positive constant C. Similarly an exponential algorithm could increase in proportion to $E_1 \times E_2^{C \times N}$ where C is any positive constant, and E_1 and E_2 can be any expressions which grow no more than polynomially in N.

A.2 Working Out the Complexity of an Algorithm

In this section we look at three ways of building new algorithms from old and show how we can derive the worst-case complexity of the new algorithm from those of the old ones.

In the following N stands for the problem size (or problem size $\times C$, for any positive constant C).

© Springer Nature Switzerland AG 2020
M. Wallace, *Building Decision Support Systems*,
https://doi.org/10.1007/978-3-030-41732-1

Table A.1 Complexity classes and Big O notation

Complexity class	Big O notation
Constant	$O(1)$
Logarithmic	$O(\log N)$
Linear	$O(N)$
Linearithmatic	$O(N \log N)$
Quadratic	$O(N^2)$
Cubic	$O(N^3)$
Polynomial	$O(N^K)$ for some constant K
Exponential	$O(2^N)$

Table A.2 Deriving complexity

Subalgorithm complexity	Number of iterations	Resulting complexity
$O(E)$	K	$O(E)$
$O(1)$	E	$O(E)$
$O(\log N)$	N	$O(N \log N)$
$O(N)$	$\log N$	$O(N \log N)$
$O(N^{K_1})$	N^{K_2}	$O(N^{K_1 + K_2})$
$O(N^K)$	2^N	2^N
$O(2^N)$	N^K	2^N

The constructions we analyse are

- Sequential composition: the new algorithm is built from two old ones by running one old algorithm after the other.
- Iteration: the new algorithm results from one old algorithm by running it a number of times (which may depend on N)
- Alternatives: the new algorithm results from two old ones by running either one or the other

For sequential composition, and for alternatives, the complexity of the new algorithm is the higher of the complexities of the two old algorithms (this follows for alternatives because we are deriving worst-case complexity).

For iteration the derivation is intuitive, though there are several cases to consider. In principle, if $O(E_1)$ is the big O expression for the complexity of the old algorithm and E_2 is the expression for the number of iterations, then the new algorithm has complexity $O(E_1 \times E_2)$. The algorithm for testing a number for divisability by 7 had a constant time subalgorithm iterated $\log N$ times, giving a complexity $O(1 \times \log N)$ which is logarithmic in N.

The Table A.2 gives the complexity class of various iterations, where K, K_1, K_2 are constants, and E is an expression matching our complexity classes.

The fifth row of the table, for example, can be used to derive the complexity of an iterative algorithm which iterates N^2 times, and whose subalgorithm, repeated on every iteration, has linear complexity $O(N^1)$. The complexity of the iterated algorithm is therefore $O(N^{1+2} = N^3)$ which is cubic.

Solutions

Exercises from Chap. 4

Worker Task Assignment

There are 12 tasks (numbered 1..12), and 6 workers (numbered 1..6). The following array of 0..1 variables

```
array [1..6,1..12] of var 0..1:Assign ;
```

represents decisions about assigning workers to tasks. $Assign[w,t] = 1$ if worker w is assigned to task t, and otherwise $Assign[w,t] = 0$.

First Challenge

Write a model which ensures at least 2 workers are assigned to each task.

Solution

```
% Parameters
int:num_workers = 6 ;
int: num_tasks = 12 ;
int: min_work = 2 ;

set of int:workers = 1..num_workers ;
set of int: tasks = 1..num_tasks ;

% Decision Variables
array [workers,tasks] of var 0..1:Assign ;
```

© Springer Nature Switzerland AG 2020
M. Wallace, *Building Decision Support Systems*,
https://doi.org/10.1007/978-3-030-41732-1

```
% Constraint
constraint forall(t in tasks)
                (sum(w in workers)(Assign[w,t]) >= min_work) ;

solve satisfy ;
```

Second Challenge

Extend your model to ensure that no worker is assigned to more than 5 tasks.

Solution

```
% Parameters
int:num_workers = 6 ;
int: num_tasks = 12 ;
int: min_work = 2 ;
int: max_work = 5 ;

set of int:workers = 1..num_workers ;
set of int: tasks = 1..num_tasks ;

% Decision Variables
array [workers,tasks] of var 0..1:Assign ;


% Constraints
constraint forall(t in tasks)
                (sum(w in workers)(Assign[w,t]) >= min_work) ;
constraint forall(w in workers)
                (sum(t in tasks)(Assign[w,t]) <= max_work) ;

solve satisfy ;
```

Third Challenge

Certain pairs of workers cannot be assigned to the same task. In the following matrix

```
array [1..6,1..6] of 0..1: bad_pair =
    [|0,1,0,1,0,0
     |1,0,0,0,1,0
     |1,0,0,1,0,0
     |0,0,1,0,1,0
     |0,0,1,0,0,0
     |0,0,0,1,0,0 |] ;
```

$bad\_pair[w1, w2] = 1$ means that workers $w1$ and $w2$ cannot be assigned to the same task.

Extend your model of the second challenge to meet the bad-pair constraint.

Solution

```
% Parameters
int:num_workers = 6 ;
int: num_tasks = 12 ;
int: min_work = 2 ;
int: max_work = 5 ;

set of int:workers = 1..num_workers ;
set of int: tasks = 1..num_tasks ;

array [1..6,1..6] of 0..1: bad_pair =
    [|0,1,0,1,0,0
     |1,0,0,0,1,0
     |1,0,0,1,0,0
     |0,0,1,0,1,0
     |0,0,1,0,0,0
     |0,0,0,1,0,0 |] ;

% Decision Variables
array [1..6,1..12] of var 0..1:Assign ;


% Constraints

constraint forall(t in tasks)
                (sum(w in workers)(Assign[w,t]) >= min_work) ;
constraint forall(w in workers)
                (sum(t in tasks)(Assign[w,t]) <= max_work) ;

constraint forall(w1,w2 in workers)
                (bad_pair[w1,w2]=1  ->
                 forall(t in tasks)(Assign[w1,t]+Assign[w2,t]
                 <= 1)
                ) ;

solve satisfy ;
```

Knapsack

You have a bag which can carry 20 kg. You have a set of things you want to bring with you, and their weights:

```
enum items = {book, jacket, washbag, computer,
            boots, alarmclock, anorak, food} ;
array [items] of int: weight = [2,4,3,8,7,1,2,6] ;
```

First Challenge

These items have a certain value to you:

```
array [items] of int: value = [6,10,8,25,22,4,5,20] ;
```

Pack the items which you can carry in your bag that bring the highest possible total value to you.

Solution

```
% Parameters
enum items = {book, jacket, washbag, computer,
              boots, alarmclock, anorak, food} ;
array [items] of int: weight = [2,4,3,8,7,1,2,6] ;
array [items] of int: value = [6,10,8,25,22,4,5,20] ;
int:bag_weight_cap = 20 ;

% Decision Variables:
array [items] of var 0..1: Selected;

% Constraints:
constraint sum(i in items)(Selected[i]*weight[i])
           <= bag_weight_cap ;

% Objective
var int: obj;
constraint obj = sum(i in items)(Selected[i]*value[i]) ;

solve maximize(obj) ;
```

Second Challenge

The knapsack also has limited space capacity, and the total volume of items it can fit inside is 2000 cm^2. Each item has not only a weight but also a volume:

```
array [items] of int: volume =
        [250, 500, 300, 250, 650, 130, 150, 600] ;
```

Fid the best solution, as for the first challenge, but the total volume of the items in the knapsack cannot be greater than the capacity of the knapsack.

Solution

```
% Parameters
enum items = {book, jacket, washbag, computer,
              boots, alarmclock, anorak, food} ;
array [items] of int: weight = [2,4,3,8,7,1,2,6] ;
array [items] of int: value = [6,10,8,25,22,4,5,20] ;
array [items] of int: volume =
        [250, 500, 300, 250, 650, 130, 150, 600] ;
int:bag_weight_cap = 20 ;
```

```
int:bag_volume_cap = 1300 ;

% Decision Variables:
array [items] of var 0..1: Selected;

% Constraints:
constraint sum(i in items)(Selected[i]*weight[i])
            <= bag_weight_cap ;
constraint sum(i in items)(Selected[i]*volume[i])
            <= bag_volume_cap ;

% Objective
var int: obj;
constraint obj = sum(i in items)(Selected[i]*value[i]) ;

solve maximize(obj) ;
```

Third Challenge

Some things are worth more in combination, and some less. Here is the additional
(or reduced) score you get for each pair:

```
array [items,items] of int: extra_value =
    [|  0,  0,  0,-5,  0,  0,  0,  0
     |  0,  0,  0,  0,  3,  0,-2,  0
     |  0,  0,  0,  0,  0,  0,  0,  0
     |-5,  0,  0,  0,  0,-2,  0,  0
     |  0,  3,  0,  0,  0,  0,  0,  0
     |  0,  0,  0,-2,  0,  0,  0,  0
     |  0,-2,  0,  0,  0,  0,  0,  0
     |  0,  0,  0,  0,  0,  0,  0,  0
    |] ;
```

If $extra\_value[i1, i2] = 3$ and if items $i1$ and $i2$ are both in your bag then the
total value of your bag is increased by 3. Naturally if $extra\_value[i1, i2] = -2$
then it is decreased by 2.

Extend your model for the first challenge to maximize the total with the modified
values.

Solution

```
% Parameters
enum items = {book, jacket, washbag, computer,
              boots, alarmclock, anorak, food} ;
array [items] of int: weight = [2,4,3,8,7,1,2,6] ;
array [items] of int: value = [6,10,8,25,22,4,5,20] ;
int:bag_weight_cap = 20 ;

array [items] of int: volume =
      [250, 500, 300, 250, 650, 130, 150, 600] ;
int:bag_volume_cap = 1300 ;
```

```
array [items,items] of int: extra_value =
    [| 0, 0, 0,-5, 0, 0, 0, 0
     | 0, 0, 0, 0, 3, 0,-2, 0
     | 0, 0, 0, 0, 0, 0, 0, 0
     |-5, 0, 0, 0, 0,-2, 0, 0
     | 0, 3, 0, 0, 0, 0, 0, 0
     | 0, 0, 0,-2, 0, 0, 0, 0
     | 0,-2, 0, 0, 0, 0, 0, 0
     | 0, 0, 0, 0, 0, 0, 0, 0
    |] ;


% Decision Variables:
array [items] of var 0..1: Selected;

% Constraints:
constraint sum(i in items)(Selected[i]*weight[i])
          <= bag_weight_cap ;
constraint sum(i in items)(Selected[i]*volume[i])
          <= bag_volume_cap ;

% Objective
var int: obj;
constraint obj =
          sum(i in items)(Selected[i]*value[i])   +
          sum(i,j in items where i>j)
             (Selected[i]*Selected[j]*extra_value[i,j]) ;

solve maximize(obj) ;
```

Exercises for Chap. 8

Trucking

There are two warehouses, A and B, each with its own truck (truckA and truckB).
Each warehouse has four customers: *A*1, *A*2, *A*3, *A*4 and *B*1, *B*2, *B*3, *B*4.
   There is a table of distances between each pair of customers and between the
warehouses and each customer.

```
Dist =
%  A     A1     A2     A3    A4     B     B1     B2     B3    B4
[|   0, 160,   150,  590, 340,  650,  725,  560,  350, 200        % A
 | 160,   0,   260,  680, 280,  650,  820,  715,  500, 150        % A1
 | 150, 260,     0,  440, 260,  520,  620,  490,  280, 150        % A2
 | 590, 680,   440,    0, 490,  240,  390,  435,  400, 550        % A3
 | 340, 280,   260,  490,   0,  660,  800,  700,  510, 140        % A4
 | 650, 650,   520,  240, 660,    0,  160,  250,  340, 670        % B
 | 725, 820,   620,  390, 800,  160,    0,  210,  380, 780        % B1
 | 560, 715,   490,  435, 700,  250,  210,    0,  215, 650        % B2
 | 350, 500,   280,  400, 510,  340,  380,  215,    0, 450        % B3
 | 200, 150,   150,  550, 140,  670,  780,  650,  450,   0 |] ;   % B4
```

Assuming the trucks have a constant speed, these can also be understood as the time needed to travel between them.

First Challenge

Truck A starts at warehouse A; it serves the customers of warehouse A; and it returns to Warehouse A. Similarly for truck B.

The challenge is to minimise the time at which the latest truck returns to its warehouse.

Solution

Assume the `Dist` data is in a file called truckdata.dzn:

```
array[1..10,1..10] of int: Dist ;

include "truckdata.dzn" ;

% This is the maximum distance of any one leg in the tour
int:maxdist = max([Dist[i,j] | i in 1..10,j in 1..10]) ;

% Decision Variables
% 1..6 are the positions in the tour, including the starting
  position
% at the warehouse and the ending position , also in the
  warehouse

% 1..5 are indices in the Dist matrix A,A1,A2,A3,A4
array [1..6] of var 1..5: RouteA ;
% 6..10 are indices in the Dist matrix B,B1,B2,B3,B4
array [1..6] of var 6..10:RouteB ;

array[1..5] of var 0..maxdist:TimeA ;
array[1..5] of var 0..maxdist:TimeB ;

% Constraints
% Start and end at warehouse A
constraint RouteA[1]=1 /\ RouteA[6]=1 ;
% Start and end at warehouse B
constraint RouteB[1]=6 /\ RouteB[6]=6 ;

include "alldifferent.mzn" ;
constraint alldifferent([RouteA[i] | i in 1..5]) ;
constraint alldifferent([RouteB[i] | i in 1..5]) ;

% The time of leg L is the distance between position L and
  position L+1
constraint forall(i in 1..5)(TimeA[i]=Dist[RouteA[i],
RouteA[i+1]]) ;
constraint forall(i in 1..5)(TimeB[i]=Dist[ RouteB[i],
RouteB[i+1]]) ;
```

```
% Objective
var int: Obj ;
constraint Obj = max([sum(TimeA),sum(TimeB)]) ;

solve minimize(Obj) ;
```

Second Challenge

Each customer has a time window within which it must be visited, given here:

```
TimeWindow =
    [|0,  2000      %A
     |800,1800   %A1
     |100,  600   %A2
     |200,1500   %A3
     |0,    500      %A4
     |0,  2000      %B
     |200,1700   %B1
     |100,1200   %B2
     |600,2500   %B3
     |300,1700   %B4
     |] ;
```

Customer $A1$ must be visited at a time T where $800 \leq T \leq 1500$ for example. The requirement is the same as the first challenge, but additionally satisfying the time window constraints.

Solution

First we add the TimeWindow data to the file truckdata.dzn.

```
include "truckdata.dzn" ;

array [1..10,1..10] of int: Dist ;
int:maxdist = max([Dist[i,j] | i in 1..10,j in 1..10]) ;
array [1..10,1..2] of int: TimeWindow ;

% Decision Variables
array [1..6] of var 1..5: RouteA ;
array [1..6] of var 6..10:RouteB ;

array[1..5] of var 0..maxdist:TimeA ;
array[1..5] of var 0..maxdist:TimeB ;

array [1..5] of var 0..5*maxdist:ArriveA ;
array [1..5] of var 0..5*maxdist:ArriveB ;

% Constraints
constraint RouteA[1]=1 /\ RouteA[6]=1 ;
constraint RouteB[1]=6 /\ RouteB[6]=6 ;
```

```
include "alldifferent.mzn" ;
constraint alldifferent([RouteA[i] | i in 1..5]) ;
constraint alldifferent([RouteB[i] | i in 1..5]) ;

constraint forall(i in 1..5)(TimeA[i]=Dist[RouteA[i],
RouteA[i+1]]) ;
constraint forall(i in 1..5)(TimeB[i]=Dist[ RouteB[i],
RouteB[i+1]]) ;

% The arrival time is the sum of the previous leg times
constraint forall(i in 1..5)(ArriveA[i] = sum([TimeA[j] | j in
1..i-1]) );
constraint forall(i in 1..5)(ArriveB[i] = sum([TimeB[j] | j in
1..i-1]) );

% Time window constraints
constraint forall(i in 1..5)
                  (ArriveA[i] >= TimeWindow[RouteA[i],1] /\
                   ArriveA[i] <= TimeWindow[RouteA[i],2] ) ;
constraint forall(i in 1..5)
                  (ArriveB[i] >= TimeWindow[RouteB[i],1] /\
                   ArriveB[i] <= TimeWindow[RouteB[i],2] );

% Objective
var int: Obj ;
constraint Obj = max([sum(TimeA),sum(TimeB)]) ;

solve minimize(Obj) ;
```

Third Challenge

Suppose Truck A can serve any customer and so can Truck B. The customers must be visited within their time windows. Truck A must still start and finish at warehouse A, and similarly Truck B must start and finish at warehouse B.

The challenge is to minimise the time at which the latest truck returns to its warehouse.

Solution

```
include "truckdata.dzn" ;

array [1..10,1..10] of int: Dist ;
int:maxdist = max([Dist[i,j] | i in 1..10,j in 1..10]) ;
array [1..10,1..2] of int: TimeWindow ;

% Decision Variables
array [1..10] of var 1..10: RouteA ;
constraint forall(i in 1..10)(RouteA[i] != 6) ;
array [1..10] of var 2..10:RouteB ;

array[1..10] of var 0..maxdist:TimeA ;
array[1..10] of var 0..maxdist:TimeB ;
```

```
array [1..10] of var 0..5*maxdist:ArriveA ;
array [1..10] of var 0..5*maxdist:ArriveB ;

% Constraints
constraint RouteA[1]=1 /\ RouteA[10]=1 ;
constraint forall(i in 1..9)(RouteA[i]=RouteA[i+1] ->
RouteA[i]=1) ;
constraint RouteB[1]=6 /\ RouteB[10]=6 ;
constraint forall(i in 1..9)(RouteB[i]=RouteB[i+1] ->
RouteB[i]=6) ;

constraint forall(i in 1..10)(exists(j in 1..10)
                  (RouteA[j]=i \/ RouteB[j]=i) );

constraint forall(i in 1..9)(TimeA[i]=Dist[RouteA[i],
RouteA[i+1]]) ;
constraint forall(i in 1..9)(TimeB[i]=Dist[ RouteB[i],
RouteB[i+1]]) ;

% The arrival time is the sum of the previous leg times
constraint forall(i in 1..10)(ArriveA[i] = sum([TimeA[j] |
j in 1..i-1]) );
constraint forall(i in 1..10)(ArriveB[i] = sum([TimeB[j] |
j in 1..i-1]) );

% Time window constraints
constraint forall(i in 1..10)
                  (ArriveA[i] >= TimeWindow[RouteA[i],1] /\
                   ArriveA[i] <= TimeWindow[RouteA[i],2] ) ;
  constraint forall(i in 1..10)
                  (ArriveB[i] >= TimeWindow[RouteB[i],1] /\
                   ArriveB[i] <= TimeWindow[RouteB[i],2] ) ;

% Objective
var int: Obj ;
constraint Obj = max([sum(TimeA),sum(TimeB)])   ;

solve minimize(Obj) ;
```

Tour Leader Assignment

A travel company has a number of tours over a season and each tour needs a tour guide. Each tour has a starting day and location, duration (days) and ending location. For the purposes of this exercise any tour guide can lead any tour, but only one at a time! Once a tour guide leads any tour he/she must be employed for the whole season (at a standard cost of 10000). When a tour guide finishes one tour and starts another there is a travel cost for going from the end location of the first tour to the start location of the next. Exactly one tour guide must be assigned to each tour.

The input is the set of tours requiring a guide, and a list of location-location travel costs. The output is a set of itineraries—one for each active tour guide—and a total cost. The parameters are specified as follows:

```
% Cost for each active tour guide
int: tour_guide_cost = 10000 ;

enum locations ;
% Each pair of locations has a travel cost recorded as an
  integer
array [locations, locations] of int: travel_cost ;

% The total number of planned tours
int: tour_ct ;
set of int: all_tours = 1..tour_ct ;
% Each tour has a start day, duration, start location and end
  location
array [all_tours] of int: tour_start;
array [all_tours] of int: tour_dur;
array [all_tours] of locations: tour_start_loc;
array [all_tours] of locations: tour_end_loc;
```

A data file giving the parameter values is:

```
% Example data for a toy problem instance
locations = {rome, paris, prague, munich, vienna, end} ;

travel_cost =
%      Rome, Paris, Prague, Munich, Vienna, End
    [| 0,     1106,  923,    699,     764,    0    % Rome
     | 1106,  0,     886,    685,     1034,   0    % Paris
     | 923,   866,   0,      300,     251,    0    % Prague
     | 699,   685,   300,    0,       355,    0    % Munich
     | 764,   1034,  251,    355,     0,      0    % Vienna
     | 0,     0,     0,      0,       0,      0    % End
    |]   ;

tour_ct = 7 ;
tour_start = [20,25,30,40,43,50,100] ;
tour_dur =   [15, 15, 15, 12, 10, 8, 0]   ;
tour_start_loc = [paris, paris, paris, paris, munich, munich,
end] ;
tour_end_loc = [rome, rome, prague, munich, vienna, munich ,
end] ;
```

The tour guide cost is only incurred for guides who lead at least one tour, and each such guide leads a sequence of tours. The guide must travel from the end location of each tour in the sequence to the start location of the next, incurring the cost given in the travel cost matrix. Naturally each tour in the sequence must end before the next one starts. Every tour must have a guide.

Challenge

The challenge is to assign a guide to every tour, minimising the cost of the guides plus their travel costs.

Solution

Put the data into the file tour-guide.dzn. The model is:

```
include "tour-guide.dzn" ;

% Cost for each active tour guide
int: tour_guide_cost = 10000 ;

enum locations ;
% Each pair of locations has a travel cost recorded as an
  integer
array [locations, locations] of int: travel_cost ;

% The total number of planned tours
int: tour_ct ;
set of int: all_tours = 1..tour_ct ;
% Each tour has a start day, duration, start location and end
  location
array [all_tours] of int: tour_start;
array [all_tours] of int: tour_dur;
array [all_tours] of locations: tour_start_loc;
array [all_tours] of locations: tour_end_loc;

% Model
%%%%%%%%%%%%%%%%%%%%%%%%%%%%%%%%%%%%%%%%%%%%

set of int:tours = 1..tour_ct-1 ;

% Decision Variables
array [tours] of var bool: first_tour ;
array [tours] of var 1..tour_ct:succ ;
var int: total_guide_cost ;

% Constraints
% All tours must be covered
constraint
    forall (t in tours)
           (first_tour[t] \/ exists(t1 in tours)(succ[t1]=t)) ;

% The guide must have time to do both a tour and its successor
constraint
    forall (t in tours)
           (tour_start[t]+tour_dur[t] <= tour_start[succ[t]]) ;

var int: total_travel_cost =
         sum(t in tours)
```

```
                (travel_cost[tour_end_loc[t],tour_start_
                loc[succ[t]]]) ;
constraint
   total_guide_cost =
        sum (t in tours) (first_tour[t]*tour_guide_cost) ;

% Objective
var int: objective;
constraint objective = total_travel_cost + total_guide_cost ;

solve minimize  objective ;
```

Exercise for Chap. 10

Tournament

Organise a league between a number of teams who should:

- each play each other twice, once at home and once away
- not play each other again within a given number of rounds
- not play at home or away three times in a row

Solution

```
% The number of teams in the league
int: teamct= 6;
% The minimum number of rounds before teams can meet again
int: notagain =4 ;

set of int:teams = 1..teamct ;
set of int: weeks = 1..(teamct-1)*2 ;

enum loc = {home, away} ;

% Decision Variables}
% Which teams play wich other teams each round
array [weeks,teams] of var teams: opp ;
% Where each team plays, each round
array [weeks,teams] of var loc: at ;

%Constraints
% Every team plays every other team twice
constraint forall(t1,t2 in teams where t1<t2)
                (exists(w1,w2 in weeks where w1<w2)
                        (opp[w1,t1]=t2 /\ opp[w2,t1]=t2)
                ) ;

% Teams play against each other once at home and once away
constraint forall(t1,t2 in teams where t1<t2, w in weeks)
                (opp[w,t1]=t2 ->
                (opp[w,t2]=t1 /\ at[w,t1] != at[w,t2]));
```

```
% Teams can't plays each other twice
% within the given number of rounds
constraint forall(t1 in teams)
                (forall(w1,w2 in weeks where w1<w2 /\
                        w2<w1+notagain)
                    (opp[w1,t1] != opp[w2,t1])
                ) ;

% Teams cannot play at home (or away) three times in a row
constraint forall(t in teams, w  in 1..(teamct-2)*2)
                (at[w,t] != at[w+1,t] \/ at[w,t]
                != at[w+2,t]) ;

 solve satisfy ;

% MiniZinc can generate output according to a specification
% - details on the MiniZinc website
output
    [ show(t)++" plays against: "++
      concat(["w"++show(w)++": "++show(opp[w,t])++", "|w in
      weeks])++
      "\n"
    | t in teams]
    ++ ["\n"] ++
    [ show(t)++" plays at: "++
      concat(["w"++show(w)++": "++show(at[w,t])++", " |w in
      weeks])++
      "\n"
    | t in teams] ;
```

References

1. S. Ahmed, A scenario decomposition algorithm for 0–1 stochastic programs. Oper. Res. Lett. **41**(6), 565–569 (2013)
2. AIMMS B.V., Aimms optimization modeling. Accessed September 2019, 2002
3. K. Akartunali, N. Boland, I.R. Evans, M.G. Wallace, H. Waterer, Airline planning benchmark problems - part i: Characterising networks and demand using limited data. Comput. Oper. Res. **40**(3), 775–792 (2013)
4. K. Akartunali, N. Boland, I.R. Evans, M.G. Wallace, H. Waterer, Airline planning benchmark problems - part ii: Passenger groups, utility and demand allocation. Comput. Oper. Res. **40**(3), 793–804 (2013)
5. S. Amborg, A. Corneil, A. Proskurowski, Complexity of finding embeddings in a k-tree. SIAM J. Discrete Math. **8**(2), 277–284 (1987)
6. T. Balyo, A. Biere, M. Iser, C. Sinz, SAT race 2015. Artif. Intell. **241**, 45–65 (2016)
7. N. Boland, I. Evans, C. Mears, T. Niven, M. Pattison, M. Wallace, H. Waterer, Rail disruption: passenger focused recovery, in *Computers in Railways XIII*, pp. 543–553 (2012)
8. A.J. Bonner, A logic for hypothetical reasoning, in *Proc. 7th Natioan Conference on Artificial Intelligence (AAAI)* (AAAI, 1988), pp. 480–484
9. G. Boole, *An Investigation of the Laws of Thought* (Cambridge University Press, 2009). Microform be downloaded at https://archive.org/stream/investigationofl00boolrich
10. K.-H. Borgwardt, *The Simplex Method. A Probabilistic Analysis*, vol. 1 of *Algorithms and Combinatorics: Study and Research Texts* (Springer, 1987)
11. D. Brélaz, New methods to color the vertices of a graph. Commun. ACM **22**(4), 251–256 (1979)
12. E.K Burke, G. Kendall (eds.), *Search Methodologies: Introductory Tutorials in Optimization and Decision Support Techniques* (Springer, 2014)
13. C.W. Choi, W. Harvey, J.H.M. Lee, P.J. Stuckey, Finite domain bounds consistency revisited, in *AI 2006: Advances in Artificial Intelligence*, vol. 4304 of *LNCS* (Springer, 2006), pp. 49–58
14. Chuffed, The chuffed CP solver. github.com/chuffed/chuffed Accessed 30-9-2019, 2019
15. G.B. Dantzig, M.N. Thapa, *Linear Programming* (Springer, 1997)
16. S. Dash, N. B. Dobbs, O. Günlük, T. J. Nowicki, G. M. Świrszcz, Lattice-free sets, multi-branch split disjunctions, and mixed-integer programming. Math. Program. **145**(1–2), 483–508 (2014)
17. R. Dechter, Bucket elimination: A unifying framework for reasoning. Artif. Intell. **113**(1), 41–85 (1999)
18. P.E. Dunne, 19th century contributions and their impact on elements of modern computers. cgi.csc.liv.ac.uk/ ped/teachadmin/histsci/htmlform/histsci_temp/av.pdf Accessed October 2019, 1996

© Springer Nature Switzerland AG 2020
M. Wallace, *Building Decision Support Systems*,
https://doi.org/10.1007/978-3-030-41732-1

19. FICO, Fico xpress mosel. community.fico.com/FICOCommunity/s/fico-xpress-mosel-mathe matical FICO Community online, 2019
20. R. Fourer, D.M. Gay, B.W. Kernighan, *AMPL: A Modeling Language for Mathematical Programming* (Cengage Learning, 2002)
21. GAMS, Gams documentation center. www.gams.com/latest/docs/ Accessed September 2019, 2019
22. Gecode, Gecode: Generic constraint development environment. www.gecode.org Accessed 30-09-2019, 2019
23. S. Geman, D. Geman, Stochastic relaxation, Gibbs distributions, and the bayesian restoration of images. IEEE Trans. Pattern Anal. Mach. Intell. **ITPAM-6**(6), 721–741 (1984)
24. D.K. Gibson, Can we banish the phantom traffic jam? www.bbc.com/autos/story/20160428 -how-ai-will-solve-traffic-part-one Accessed 6th November 2019, 2016
25. I. Gilboa, A.W. Postlethwaite, D. Schmeidler, Probability and uncertainty in economic modelling. J. Econ. Perspect. **22**(3), 173–188 (2008)
26. P.E. Gill, W. Murray, A numerically stable form of the simplex algorithm. Linear Algebra Appl. **7**(2), 99–138 (1973)
27. R. Gomory, *All Integer Programming Algorithm* (Prentice-Hall, 1963), pp. 193–206
28. L. Granvilliers, F. Benhamou, Algorithm 852: Realpaver: an interval solver using constraint satisfaction techniques. ACM Trans. Math. Softw. **32**, 138–156 (2006)
29. Guardian Cities and A Gross, The four hour commute: the punishing grind of life on Sao Paulo's periphery. www.theguardian.com/cities/2017/nov/29/four-hour-commute-grind-life-sao-paulo-periphery Accessed 5 November 2019, 2017
30. D. Hemmi, G. Tack, M.G. Wallace, A recursive scenario decomposition algorithm for combinatorial multistage stochastic optimisation problems, in *Proceedings of the Thirty-Second AAAI Conference on Artificial Intelligence (AAAI-18)* (AAAI, 2018), pp. 1322–1329
31. IBM, OPL, the modeling language. www.ibm.com/support/knowledgecenter/SSSA5P_12.7.0/ ilog.odms.ide.help/OPL_Studio/opllangref/topics/opl_langref_modeling_language.html IBM Knowledge Centre, 2019
32. P. Jegou, H. Kanso, C. Terrioux, On the relevance of optimal tree decompositions for constraint networks, in *IEEE 30th International Conference on Tools with Artificial Intelligence (ICTAI)*, pp. 738–743 (2018)
33. N. Karmarkar, A new polynomial-time algorithm for linear programming. Combinatorica **4**(4), 373–395 (1984)
34. V. Klee, G.J. Minty, How good is the simplex algorithm? in *Proc. 3rd Symposium on Inequalities* (Academic Press, 1972), pp. 159–175
35. A.H. Land, A.G. Doig, An automatic method of solving discrete programming problems. Econometrica **28**(3), 497–520 (1960)
36. C. Lecoutre, *Constraint Networks: Targeting Simplicity for Techniques and Algorithms* (Wiley, 2013)
37. R. Lewis, Metaheuristics can solve Sudoku puzzles. J. Heuristics **13**(4), 387–401 (2007)
38. H. Lo, S. Blumsack, P. Hines, S. Meyn, Electricity rates for the zero marginal cost grid. Electr. J. **32**(3), 39–43 (2019)
39. M. Luby, A. Sinclair, D. Zuckerman, Optimal speedup of Las Vegas algorithms. Inf. Process. Lett. **47**(4), 173–180 (1993)
40. J. Luedtke, S. Ahmed, A sample approximation approach for optimization with probabilistic constraints. SIAM J. Optim. **19**(2), 674–699 (2008)
41. A.K. Mackworth, Consistency in networks of relations, in *Readings in Artificial Intelligence*, ed. By B.L. Webber, N.J. Nilsson (Morgan Kaufmann, 1981), pp. 69–78
42. J. Marques-Silva, K.A. Sakallah, GRASP—A new search algorithm for satisfiability, in *Proc. ICCAD '96* (IEEE Computer Society, 1996), pp. 220–227
43. R. Martí, P. Panos, M. Resende, *Handbook of Heuristics*. Handbook of Heuristics (Springer, 2017)
44. E. Mendelson, *Introduction to Mathematical Logic* (Van Nostrand, 1964)

45. R.T. Milam, M. Birnbaum, C. Ganson, S. Handy, J. Walters, Closing the induced vehicle travel gap between research and practice. Transp. Res. Rec. **2653**, 10–16 (2017)
46. MiniZinc, Basic search annotations. www.minizinc.org/doc-2.3.1/en/mzn_search.html#search-annotations Accessed 30-9-2019, 2019
47. MiniZinc, Solving technologies and solver backends. www.minizinc.org/doc-2.3.1/en/solvers.html Accessed 30-9-2019, 2019
48. M. W. Moskewicz, C. F. Madigan, Y. Zhao, L. Zhang, S. Malik, Chaff: Engineering an efficient SAT solver, in *Proceedings of the 38th Annual Design Automation Conference*, DAC '01 (ACM, 2001), pp. 530–535
49. J.F. Muth, G.L. Thompson, *Industrial Scheduling* (Prentice-Hall, 1963)
50. Opturion, Case studies. opturion.com/production-scheduling Accessed October 2019, 2019
51. J. Rintanen, *Planning as Satisfiability* (IOS Press, 2009), pp. 483–504
52. S. Ropke, D. Pisinger, An adaptive large neighborhood search heuristic for the pickup and delivery problem with time windows. Transp. Sci. **40**, 455–472 (2006)
53. M. J. Saltzman, *Coin-Or: An Open-Source Library for Optimization* (Springer US, 2002), pp. 3–32
54. C. Schulte, P.J. Stuckey, When do bounds and domain propagation lead to the same search space? ACM Trans. Program. Lang. Syst. **27**, 388–425 (2005)
55. H. Simonis, Sudoku as a constraint problem, in *Proc. 4th Int. Workshop on Modelling and Reformulating Constraint SatisfactionProblems*, pp. 13–27 (2005)
56. S. Singh, *Fermat's Last Theorem* (HarperCollins, 2012)
57. J. Stangroom, *Einstein's Riddle: Riddles, Paradoxes, and Conundrums to Stretch Your Mind* (Bloomsbury, 2009)
58. P. J. Stuckey, Lazy clause generation: Combining the power of SAT and CP (and MIP?) solving, in *Integration of AI and OR Techniques in Constraint Programming for Combinatorial Optimization Problems*, ed. By A. Lodi, M. Milano, P. Toth (Springer, Berlin, Heidelberg, 2010), pp. 5–9
59. P.J. Stuckey, J.H.M. Lee, Advanced modeling for discrete optimization. www.coursera.org/learn/advanced-modeling Accessed September 2019, 2019
60. P.J. Stuckey, J.H.M. Lee, Basic modeling for discrete optimization. www.coursera.org/learn/basic-modeling Accessed September 2019, 2019
61. P.J. Stuckey, K. Marriott, G. Tack, The MiniZinc handbook. www.minizinc.org/doc-2.3.1/en/index.html Accessed 30-9-2019, 2019
62. Sudoku, Sudoku 4484: Level diabolical. *The Age Newspaper*, p. 30, 2019
63. N. Tamura, T. Tanjo, M. Banbara, Solving constraint satisfaction problems with sat technology, in *Functional and Logic Programming* (Springer, Berlin, Heidelberg, 2010), pp. 19–23
64. P. Van Hentenryck, L. Michel, *Numerica: A Modeling Language for Global Optimization* (MIT Press, 1997)
65. D. Wedelin, Optimization models. www.cse.chalmers.se/edu/year/2010/course/ DAT026/CourseMaterial/opt2Intro-2014.pdf Course Material, 2003
66. Wikipedia, Coin-or. https://en.wikipedia.org/wiki/COIN-OR Accessed on 30-09-2019, 2019
67. H.P. Williams, *Model Building in Mathematical Programming* (Wiley, 2013). Accessed September 2019
68. E. Winsberg, *Philosophy and Climate Science* (Cambridge University Press, 2018)

Index

Printed in the United States
By Bookmasters